Viruses and the Environment

To Audrey, Wendy and Elaine

Viruses and the Environment

Second edition

J.I. Cooper

Natural Environmental Research Council, Institute of Virology and
Environmental Microbiology, Oxford, UK

CHAPMAN & HALL

London · Glasgow · Weinheim · New York · Tokyo · Melbourne · Madras

Published by Chapman & Hall, 2–6 Boundary Row, London SE1 8HN, UK

Chapman & Hall, 2–6 Boundary Row, London SE1 8HN, UK

Blackie Academic & Professional, Wester Cleddens Road, Bishobriggs, Glasgow G64 2NZ, UK

Chapman & Hall GmbH, Pappelallee 3, 69469 Weinheim, Germany

Chapman & Hall USA, 115 Fifth Avenue, New York, NY 10003, USA

Chapman & Hall Japan, ITP-Japan, Kyowa Building, 3F, 2-2-1 Hirakawacho, Chiyoda-ku, Tokyo 102, Japan

Chapman & Hall Australia, 102 Dodds Street, South Melbourne, Victoria 3205, Australia

Chapman & Hall India, R. Seshadri, 32 Second Main Road, CIT East, Madras 600 035, India

First edition 1984

Second edition 1995

© 1984 J.I. Cooper and F.O. MacCallum

© 1995 J.I. Cooper

Typeset in 10/12 Palatino by Best-set Typesetter Ltd., Hong Kong
Printed in Great Britain at the University Press, Cambridge

ISBN 0 412 45120 4

A catalogue record for this book is available from the British Library

Library of Congress Catalog Card Number: 95-71232

∞ Printed on permanent acid-free text paper, manufactured in accordance with ANSI/NISO Z39.48-1992 and ANSI/NISO Z39.48-1984 (Permanence of Paper).

Contents

Preface

In the decade since the first edition of this book was published there has been unprecedented public interest in viruses. On a world scale, one of the greatest driving forces was the current pandemic of human immunodeficiency lentiviruses (HIVs). More locally, the many mysteries surrounding infectious dementias such as that manifested in cattle and which acquired the name 'mad cow disease' attracted interest. This was largely because of fears that people were at risk from a terrifyingly fatal illness caused by an agent in their food. The still uncharacterized agents were strongly suspected to have crossed species boundaries (from sheep to cattle and zoo animals) and suspicion surrounded the use of therapeutic products derived from animal carcasses because exposure to such animal extracts correlated with dementias in humans. Such fears justified commercial penalties on animal movement and triggered revision of systems for meat processing and these directly impacted on farm incomes in the United Kingdom. Other localized loss occurred in pigs because of a hitherto unrecorded syndrome (blue ear disease) caused by a corona-like agent 'Lelystad virus' (Meulenberg *et al.*, 1993). There were also incidents which were treated as 'seven-day wonders' by the print media – as with the occurrence of measles-like morbilliviruses in marine mammals or horses and the hitherto unnoticed human respiratory syndrome first noticed and linked to enhanced rodent food supply (tree seed) which encouraged the build-up of hantavirus isolates which were imaginatively called 'Sin Nombre' (= no name). Other virological phenomena attracted fewer column inches than they warranted. Thus, the fact that a parvovirus mutated and that the mutant spread worldwide in as little as 2 years was not widely publicized outside the specialized scientific literature (about the same time, the calicivirus causing haemorrhagic disease in rabbits became established worldwide – possibly as a result of dissemination in frozen rabbit from the Peoples' Republic of China; Chasey, 1994). In a general sense, dog owners were made aware that their puppies needed to be given protective inoculations against a virus other than that which causes distemper but the catastrophic impact of the mutant parvovirus on rare or elusive wildlife was not widely publicized.

By contrast, a few viruses have been given extreme publicity – perhaps because they kill in a particularly dramatic fashion and have the capacity for spreading beyond the remoteness of Africa where they were first recognized. The potential threat to humans from viruses circulating silently in unknown wildlife had long been known in the context of flaviviruses and rhabdoviruses associated with tropical fevers or rabies, but the publication during 1994 of two semi-popular books, *The Coming Plague, Newly Emerging Diseases in a World out of Balance*, by Laurie Garrett, and *The Hot Zone*, by R. Preston, heightened awareness of filoviruses and the 1995 epidemic of Ebola haemorrhagic fever in Zaire re-emphasized the dangers.

While the unexpected abundance of bovine spongiform encephal-opathy and speculation about the origin and composition of the aetiological agent kept the pot boiling (particularly in Europe), other virological events occasioned less comment than they deserved. At least in developed countries of the world, where the associated disease was already a distant memory, the probable global eradication of smallpox and the real possibility of measles eradication have been eclipsed by the more immediate hazard from HIVs and their sequelae (Spencer and Price, 1992). Despite dramatic amounts of coordinated research investment, many questions concerning lentiviruses remain to be answered. Nevertheless, because education was seen to provide one of the few immediately worthwhile ways of managing this disease, the hitherto arcane and often private experiences (including conflicts) of specialized research scientists have been brought into the public domain via both the film and print media.

As a result, there is now much more general knowledge about what viruses are, how they spread (particularly venereally or in association with blood) and how rapidly they can change.

The past decade has seen routine (even automated) nucleic acid characterization systems and the transfer of DNA between a variety of lifeforms has become commonplace, with bacterial viruses playing crucial roles in these processes. An early application of the technology was the 'engineering' of crop plants and, partly because of their small size and presumed simplicity, viruses were among the first sources of transgenic genes. In a few instances, the concept of gene transfer attracted strong reactions (riotous assembly and even bomb blast). However, the genetic engineering of microbes (particularly for contained use by industry) attracted only transient interest and the use of viruses to remedy metabolic deficiencies in humans or for vaccine delivery attracted little, if any, critical comment. Indeed, the reality that people containing genetically engineered viruses circulate freely in urban communities attracted no general comment at all. Paradoxically, the production of transgenic farm animals has been attacked as unethical and unnecessary and the imminent commercialization of food crops containing virus-derived transgenic tolerance genes is a matter of concern to people from many backgrounds. Commercial interests see financial opportunities, real benefits and essentially no risk in such technologies. Possible negative impact on wildlife is sometimes trivialized as fantasy. However, this attitude is unlikely to be helpful. The lay public are most likely to be receptive of the new technologies if they have been made aware of the background and, perhaps most important, the benefits which are expected. Medical need is readily assimilated and the contained use of genetically engineered microorganisms in industrial processes is not perceived as threatening. However, debate on the 'planned release' of genetically modified organisms into the environment has barely started and is certainly not yet resolved although there have been hundreds of authorized experimental plantings without obvious harm. Interestingly, the prospect of large-scale plantations of novel plant genotypes deliberately manufactured to contain virus-derived components has highlighted ignorance concerning the role played by viruses in natural communities. Questions which now need answers are: do viruses really regulate (wild) plant survival; what are the interactions with herbivores; and to what extent will increased planting of transgenic crops enhance opportunities for the evolution of new viral pathotypes? Analogous questions relate to insect pathogenic viruses which in essentially native form were in the vanguard of biological control agents for field (particularly forest) use. Now that attempts are being made to enhance the effectiveness of these insect pathogenic viruses, awareness has increased about whether there is real potential for unexpected environmental damage (by lessening first insects and thereafter the vertebrates which exploit this resource).

When the first edition of this book was published, recombination in RNA virus genomes had been recognized but the process seemed unlikely to be more than a curiosity. Subsequently, computer-aided genome sequence comparisons provided retrospective assessments of viral evolution and strongly suggested that modular exchange of pre-developed functional units (virus genes) is a regular means of rapid evolution in at least some categories of viruses. The analyses revealed unexpected diversity among viral genomes and the implications for virus taxonomy have not yet been fully addressed. It is now known that viral genomes are chimeral and it is also becoming apparent that recombination is a normal cause of virus variation. Consequently, even the most up-to-date official views about viral relationships which have been used in this book are manifestly flawed because the taxonomies are largely based on one or two traits rather than the whole plastic context in which those traits are embedded.

Regrettably, Fred MacCallum, the co-author of the first edition of this book, died in September, 1994. Fred's working life embraced an unsurpassed range of viruses and his contributions were both numerous and diverse. His principal work concerned viruses in the context of human disease. However, at the start of his research career, he was part of the team (with F.C. Bawden and K.M. Smith) who collaborated with Radcliffe Salaman in the characterization of potato viruses. Fred brought serological tools to bear on these perplexing plant pathogens which had some properties in common with the rickettsia-like agents he studied in the context of human disease. Towards the end of his career, Fred was concerned with safety aspects of insect pathogenic baculoviruses and, after retirement from hospital practice, he concentrated more and more on comparative aspects of virology and the implications of his finding in widely scattered humans, domesticated and wild animals of antibodies which reacted with the small RNA virus from *Darna trima*: a lepidopterous pest of oil palms.

Acknowledgements

I am indebted to Mrs Stephanie Price for cheerfully typing the text and grateful to the authors and publishers who gave permission to reproduce illustrations.

The nature of viruses 1

1.1 INTRODUCTION

Viruses are ubiquitous, yet when plants, animals and other biota are in their natural environments, it is commonplace for virus replication to occur in hosts that show no detectable abnormality. In many instances, these inapparent (sometimes called latent or silent) infections are potentially hazardous for humans, their domestic animals or crops. From time to time, drastic environmental changes occur and whether these are attributable to natural phenomena or result from human interference (agriculture, forestry, urbanization, war) one consequence is liable to be a catastrophic increase in disease incidence (epidemics). Analyses of epidemics have furthered understanding of the complex interplay of factors that modify host susceptibility/sensitivity on the one hand, while influencing the abundance, virulence and invasiveness of the viruses on the other. In recent years a variable of unknown magnitude has been introduced into the environment. Diverse genetically engineered crops and microorganisms now contain virus-derived sequences and, although their distribution is presently regulated and contained, there are strong pressures for commercialization and more widespread use. Considerable scope exists for speculation about the epidemiological impact of virus-derived transgenic genes. Experiments designed to address the limits on gene pools available to viruses, the time scale for virus evolution and the consequences will give further insight into the private lives of virus populations that wax and wane in response to the constant fluctuations in their microenvironments – out of sight and usually out of mind. Although considerable ignorance exists, the unmanaged components of the environment commonly are reservoirs of virus biodiversity – wild plants, including weeds in relation to agricultural/horticultural crops, migratory birds or commensal rodents for human infections. Invertebrates, usually arthropods, are the omnipresent go-betweens. However, virus transmission may be achieved independently by contact or when the virus infects reproductive organs, e.g. plant seed and pollen, fungal spores, animal spermatozoa and ova. Alternatively, or additionally, the genetic constituents (genomes) of viruses may become intimately associated and replicate in harmony with the genomes of their hosts. There is a mass of knowledge concerning viruses in the environment. In this small book I have attempted only to whet the appetite of readers by providing a general introduction to the concepts and to the literature. Many of the terms in this book have been defined when first used; most have also been described in the dictionary by Rowson et al. (1981) or in Black's Medical or Veterinary Dictionaries (e.g. West, 1995). The multi-authored (multiple volume) works edited by Fields *et al.* (1990) or Webster and Granoff (1994) provide the most comprehensive resources. The occasional series entitled *Seminars in Virology* and the review periodical *Advances in Virus Research* provide a wealth of complementary information.

1.2 VIRUS CHARACTERISTICS

The ability of viruses to function as parasites within cells of their hosts is shared with microorganisms, including for example fungi, algae, bacteria and protozoa. However, viruses have some features that are unique: their nature requires that parasitism occurs at the genetic level and not by crude competition for space and nutriment. In essence, viruses are nucleic acids which can copy themselves when in an appropriate cell. To facilitate their dispersal, viruses generally need protection and, with few exceptions, viruses are composed of morphologically distinctive particles (virions) having two main components: a nucleic acid genome and a proteinaceous coat (capsid) that may have more than one distinguishable polypeptide sequence. Some viruses are characterized by the presence within the virions of enzymes, carbohydrates and/or lipids but because they use only one type of nucleic acid (either DNA or RNA), viruses differ fundamentally from microorganisms.

Viruses constitute a class of obligate parasites that is very heterogeneous but (as indicated in Figure 1.1) characterized by dimensions measured in nanometres (nm). En masse, virions sometimes aggregate and much knowledge about their structure has been derived from X-ray crystallography: the morphology of individual particles is revealed routinely by electron microscopy (now refined by use of cryo-techniques). Figure 1.2 indicates the shape and relative sizes of representative virus taxa. Most virions are roughly spherical, some are more or less rigid cylinders and a few have complex shapes or are pleomorphic (i.e. their shape varies).

The spectrum of agents qualifying for description as viruses extends

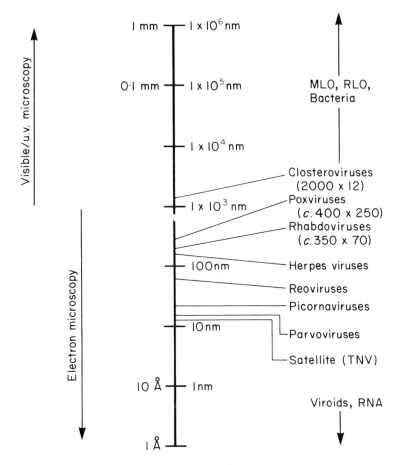

Figure 1.1 The size spectrum of virus and virus-like agents. MLO, mycoplasma-like organisms; RLO, rickettsia-like organisms; TNV, tobacco necrosis virus.

into the realm of nucleic acids; in many instances the ultimate infectious unit is not a nucleoprotein virion but a nucleic acid that can directly facilitate the synthesis of polypeptides on ribosomes of host cells. The first viruses of this type to be recognized were plant pathogens in which the essential genetic information (genome) was in one piece (Gierer and Schramm, 1956). This is now known to be exceptional in two respects. Numerous viruses have virions that contain nucleic acids lacking the property of direct infectiousness; the genomic message requires copying before translation into proteins on ribosomes and is described as 'negative sense' or having 'negative polarity'. An inter-mediate condition ('ambisense genomes') has also been recognised, but seems to be an uncommon use of genetic resources (Bishop, 1986). Additionally, a diverse array of viruses possess genomes composed of

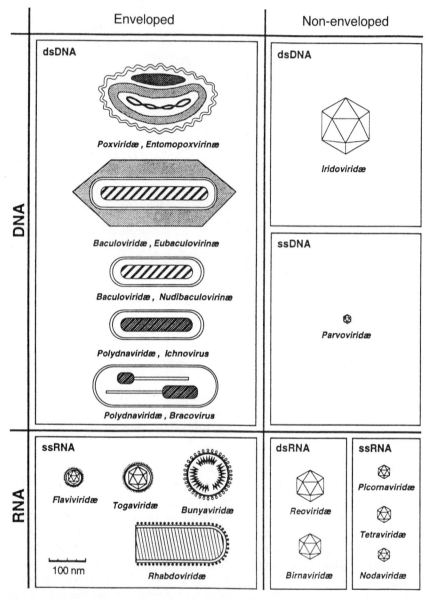

FAMILIES OF VIRUSES INFECTING INVERTEBRATES

Figure 1.2 Diagrams at the same scale to indicate the shapes and relative sizes of viruses. Reproduced with permission from Springer-Verlag.

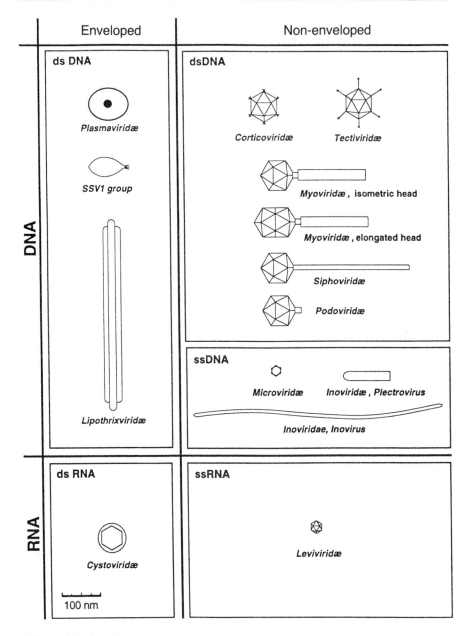

Figure 1.2 *Continued*

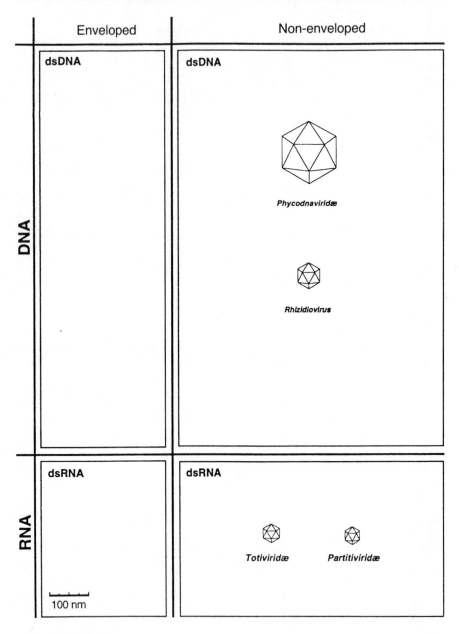

Figure 1.2 *Continued*

FAMILIES OF VIRUSES INFECTING VERTEBRATES

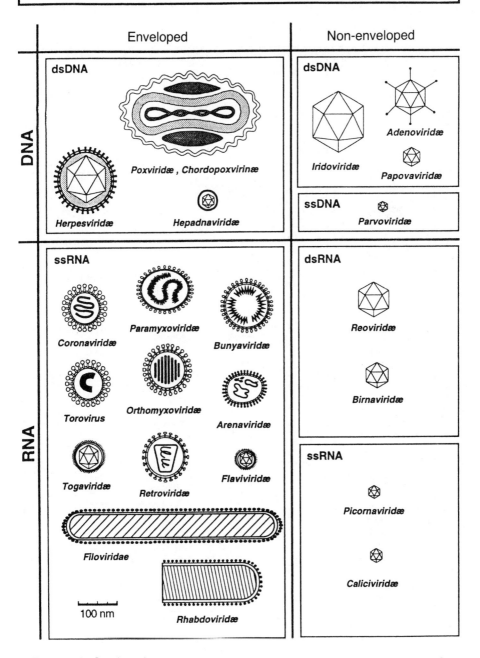

Figure 1.2 *Continued*

FAMILIES AND GROUPS OF VIRUSES INFECTING PLANTS

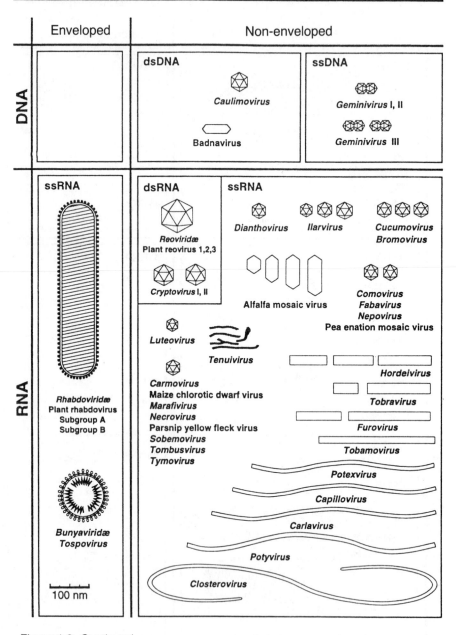

Figure 1.2 *Continued*

two, three or more separate nucleic acid species (see p. 28). In some instances the differing parts of these divided genomes occur together in each virion but in others the constituent parts are encapsidated (packaged) separately. Alfalfa mosaic ilarvirus indicates the complexity: the virions are characterized by RNA of four lengths separately packaged in particles differing both in size and shape. With this virus, the three largest nucleic acid molecules (conventionally described as RNA-1, RNA-2 and RNA-3 in decreasing size order) comprise the genome; the RNA-4 is a subgenomic messenger coding for the capsid protein. The three genomic RNA pieces, plus either the fourth RNA or the capsid protein, are needed for complete expression of infectivity. In this context, it is noteworthy that some viruses encapsidate extra-genomic RNA segments that are at times host-derived or fragments of virus-coded RNA. Interestingly, these extragenomic RNAs sometimes modify the disease severity (pathogenicity) determined by the 'normal' virus genome. In viruses with multipartite genomes, the constituent RNAs are not always easy to classify because there is facultative 'use' of the available genetic resources. Thus, beet necrotic yellow vein furovirus, when isolated from roots of naturally infected sugar beet, contains four or five distinguishable RNAs. During experimental maintenance of the virus through many subcultures, the third, fourth and fifth RNAs are deleted or lost from the population and it is thereby arguable that only two RNA species are 'the genome' of this virus (Brunt and Richards, 1989). Roles for the facultative RNA species are not fully elucidated, but one facilitates dissemination of the virus between cells in sugar beet roots and another is essential for efficient transmission by fungal vectors.

The plasticity of a viral genome and the opportunities which exist for stable relationships to develop from casual associations give rise to problems in definition and nomenclature which were addressed by Liu and Cooper (1994) – in the context of 'satellitism'. As the sensitivity of nucleic acid detection and characterization methods increased, it became apparent that extragenomic nucleic acids of many types are encapsidated within virions of diverse taxonomic affinities.

In sexually reproducing eukaryotes, a species is defined in terms of individuals sharing in a common gene pool. Sequence analyses show that viruses also share genes and are not purely asexual entities. However, the degree to which viruses are genetically isolated from one another is unknown. If gene exchanges are frequent, the concept of a viral species is difficult to apply. The continuum of nucleic acids encapsidated by viruses probably results from recombination. Copying errors are facilitated even by casual associations between viral components but become more important when associations have been stabilized by selection and mutual benefit.

Re-assortment, resulting in new virus genotypes (and phenotypes) is facilitated when viral genomes are divided between different particles. However, there is also a risk that chance will separate the components and, in isolation, these nucleic acids are usually unable to replicate. However, examples are known in which only one part of a genome can survive (e.g. the RNA-1 of tobraviruses in potato). This RNA species does not encode coat protein and is detected because it is associated with disease. Other capsid-deficient nucleic acids (umbraviruses) can be experimentally maintained in isolation but are not known to occur naturally except in association with another virus (luteovirus) which encodes a capsid protein which both genomes share.

1.3 AGENTS THAT WERE CONFUSED WITH VIRUSES AND HOW THEY DIFFER

Classically, the presence of viruses was inferred if extracts of diseased organisms that had passed through filters retaining bacteria still contained infectious pathogenic agents. Now, however, four additional classes of filter-passing pathogen are known.

New chapters in the history of pathology were opened when a distinct group of intracellular parasites (Bedsonia/Chlamydia) associated with diseases (psittacosis, lymphogranuloma venereum, trachoma) or latent infection in birds and mammals was recognized as distinct from viruses (Moulder, 1966; Stortz and Page, 1971). In plant pathology confusion persisted longer but, since 1967, three main groups of antibiotic-sensitive microorganisms have been distinguished from viruses (Hopkins, 1977; Purcell, 1979; Whitcomb and Tully, 1979). In common with some viruses, the spread of these microorganisms is facilitated by the feeding of insects but whereas symptoms attributable to viruses are unaffected, diseases associated with the microorganisms classified as spiroplasmas or mollicutes (*sensu* Freundt, 1974) are suppressed by tetracycline antibiotics. Rickettsia-like organisms are distinguishable in being sensitive to penicillins which do not seem to affect either viruses or mollicutes.

A class of single-stranded RNA molecules (which may be linear or covalently closed rings) have such a unique composition that they seem unlikely to have either derived from or given rise to viruses that are currently recognized. These nucleic acids, which do not code for proteins, form another extension of the size spectrum of virus-like agents and are generally described as viroids (Diener, 1979). Several viroids have been cloned, sequenced and transcribed *in vitro*. Despite their small size (250–350 bases) and propensity to re-arrangement, viroids are unambiguously pathogenic. Associated diseases include potato spindle tuber, which lessens potato yield by 20–70%, and cadang-

cadang of coconuts, which has been estimated to affect 30 million trees valued at $100 million. It was earlier suggested that viroids are aberrant RNA components normally having regulatory functions in plant cells, but the possibility cannot be excluded that viroids also exist within animal cells. Indeed, Elena *et al.* (1991) showed a common phylogeny for viroids and the hepatitis delta agent (hepadnavirus) which they interpreted as indicating that these molecules are fossils from a pre-cellular world (see also Gesteland and Atkins, 1993). Parallels have been drawn between viroids known from plants and the agents of transmissible dementias which affect sheep and goats, zoo animals, and wildlife including mink and bovines. The same agents are implicated with Kuru-Kuru or Jakob – Creutzfeldt disease in humans (Lacey, 1994). Evidence of similarity between viroids and these agents (termed prions; Prusiner, 1992) is sparse and inconclusive. Indeed, prions are currently considered to be proteins which modify the activities of enzymes rather than nucleic acids.

1.4 VIRUS NAMES

Traditionally, diseases were recognized and, when other causes had been excluded, the diseases were attributed to virus-like agents that acquired vernacular names which described the symptoms, the host or the geographical location where they were first noticed. Virus names generated in this way tend to be cumbersome but nevertheless provide useful information. An alternative approach has been to assign isolates number- or letter-based codes. The appropriateness of these schema was called into question with the routine isolation from vertebrates or plants of 'orphan' (in search of a disease) or 'cryptic' viruses and the observation that one virus can infect more than one species of host in which several viruses might coexist. The frequency of multiple infection and the degree of confidence in the numbers of distinct agents present in any individual depends on the sensitivity and specificity of tools used. Nevertheless, when seeking to establish causal associations, it is necessary to identify unambiguously a virus and to distinguish one agent from others that might also be present. Unfortunately, this is not always technically feasible. Even when possible, the identification of the causal agent of disease is not routinely done because of the time scale before symptom expression or for ethical reasons, as when the infection has fatal consequences and the hosts are human. Because knowledge is incomplete, synonymy is prevalent but diminishing. Numerous attempts have been made to harmonize and rationalize systems of virus naming (Bergold *et al.*, 1960; Gibbs, 1969; Hansen, 1975; Williams and Cory, 1994). To facilitate comparisons and prediction, guidelines have been published for the identification of plant pathogenic (Hamilton *et al.*, 1981) and bacterial viruses (Ackermann *et al.*, 1978). Nevertheless, routine identifications tend to be based on one property that is conveniently measured, even though that feature may relate to only a small part of the genome. Thus, undue taxonomic weight is often given to the source host, the antigenic properties of virions or vector specificities. The molecular basis of vector relations is not understood in most systems but may, at an extreme, derive from three amino acids (e.g. DAG, NAG, VQV; Harrison and Robinson, 1988; Kreiah *et al.*, 1994) – determined by as little as one ten-thousandth of the total coding capacity. Virus – vector determination is probably multifaceted; in potyviruses two genome-encoded proteins define aphid transmissibility (Atreya *et al.*, 1990). Interestingly, deficiencies in one trait can be complemented *in trans* (during co-infection by another potyvirus with a functional trait; Hobbs and McLaughlin, 1990). Notwithstanding these complications, a systematic classification of viruses is evolving.

1.5 VIRUS CLASSIFICATION

The latinized binomial nomenclature in which pairs of names are used to describe plants (e.g. *Cucumis sativus*, cucumber) or animals (e.g. *Vulpes vulpes*, red fox) is a well-defined biological objective which serves to identify genetically distinct units of diversity. Depending upon the number of characteristics available for use, this categorization is necessarily achieved with an element of subjectivity. While the concept of a species (identified by the second name of the pair) recognizes distinctness, the genus (the first part of the binomial) indicates degree of association, similarity, or perhaps relationship. In line with this concept, the diversity of bacterial viruses (bacteriophages) and the frequency of morphological descriptions is being rationalized; Ackermann (1992) grouped 3850 recorded phages (*c.* 95% of those recognized) into one order (Caudivirales) embracing three families with double-stranded (ds) linear DNA genomes: Myoviridae, Siphoviridae and Podoviridae. Analogously, genera of viruses with monopartite negative-sense RNA genomes have been grouped into families and classes within the order Mononegavirales (Pringle, 1991).

The European flora and fauna which were first categorized in this way (by Linnaeus, 1707–1778) had been studied or at least recognized for generations before. By contrast, even though there has been a recent burgeoning of base sequence data and increasing accessibility of computer-based systems for managing and collating such data, little information about most viruses is presently available. Current knowledge indicates that viruses *in vivo* are by no means genetically fixed; populations are polymorphic to an extent that is almost impossible to measure reproducibly and heterozygous to a hitherto unexpected degree. Furthermore, intramolecular recombination is not uncommon. Nevertheless, based on experiences with a modest number of model systems which may or may not be representative of real virus populations, the International Committee on Taxonomy of Viruses (ICTV; under the aegis of the Virology Section of the Union of Microbiological Societies) has recognized a diverse array of family and genus names (see Appendix 1.1).

The ICTV-authorized classification system was not intended to have evolutionary significance. Nevertheless, patterns of virus evolution have been inferred from comparison of base sequences and relationships among viruses from diverse hosts are now accepted (albeit with reservations in some instances). The first virus species and names for them were provisionally approved in 1981 and a substantial number of additional taxa were accepted in 1994. Unfortunately, some viral genera and families were established in the absence of detailed sequence information and before recognition of the composite nature of their genomes. Consequently, some of the family and genus names currently approved by ICTV (Murphy *et al.*, 1995) may not gain wide acceptance. Virologists concerned largely with plant pathogens resisted using the well-established descriptors, e.g. family names ending 'viridae', but recognized genus and family names have been publicized (Mayo and Martelli, 1993). Structural and organizational similarities between viruses infecting plants or animals have been recognized for several years (Kamer and Argos, 1984; Bruenn, 1991; Goldbach *et al.*, 1991) but the validity of 'higher' groupings (often designated 'supergroups' rather than Orders) is actively debated. Notwithstanding differences in genome structure (Zaccomer *et al.*, 1995), translation strategies, capsid morphology, vectors and hosts, it is undoubtedly possible to trace partial sequences with recognizable features in common. Plausibly, these features indicate the scope of gene pools and evolutionary origins (see Zanotti *et al.*, 1995). In any event the chimeral nature of viruses is now becoming abundantly clear.

In this book, the ICTV-approved names are generally used but, because of its descriptiveness and relevance in an ecological context, the term arbovirus (deriving from *arthropod-borne*) also appears. Arboviruses were defined by the World Health Organization as 'viruses which are maintained principally or to an important extent through biological transmission between vertebrate hosts by haematophagous arthropods; they multiply and produce viraemia (occur in blood) in the vertebrates, multiply in tissues of the arthropods and are passed on to new vertebrates by the bites of arthropods (in which they replicate) after a period of extrinsic incubation'. More than 200 viruses in six families approved by ICTV (Flaviviridae, Togaviridae, Bunyaviridae,

Iridoviridae, Rhabdoviridae and Reoviridae) are included with a similar number of unclassified probable viruses (Monath, 1986). It is noteworthy that the Togaviridae is a family of viruses having members which are not arthropod-borne, and very few of the known viruses in the Iridoviridae have been associated with vertebrates. Furthermore, some of the viruses (genus *Tospovirus*) in the Bunyaviridae replicate in plants and also in their thysanopteran (thrips) vectors.

The detailed properties of many individual viruses and known synonyms have been collated by numerous individuals and groups (Martyn, 1971; Andrews *et al.*, 1978; Ackermann, 1992) but the sixth report of the International Committee on Taxonomy of Viruses (Murphy *et al.*, 1995) is the most up-to-date conspectus of information. The following paragraphs outline a few of the biological properties of some ICTV-approved taxa that were selected to indicate either the financial and sociological burden attributable or other noteworthy attributes such as the diversity of known hosts. Where appropriate, both the colloquial and the scientific names of virus-associated diseases have been given together. Although Family and Genus names should formally be printed in italics (or underlined), I have not rigorously followed that rule while generally following the taxonomic hierarchy proposed by Murphy *et al.* (1995).

1.5.1 SOME APPROVED VIRUS TAXA

Poxvirideae

Viruses assigned to this family have dsDNA genomes and have been recognized in vertebrate animals, birds and insects. The epidemiology of entomopoxviruses (infecting invertebrates) is largely unknown but poxviruses affecting vertebrates have been studied for many years. Short-range spread is via aerosols or biting flies and poxviruses from one vertebrate species tend to infect few others: the associated symptoms of disease are pustules that tend to be localized and transient. However, other poxviruses produce tumours (e.g. myxoma virus causing myxomatosis in wild and domesticated European rabbits, *Oryctolagus* spp.) or generalized rashes associated with frequent mortality. In 1520, the Aztec nation is thought to have lost half of its (3.5 million) population to smallpox (variola). Currently, only two laboratories are known to hold infectious cultures of the virus. Since a comprehensive range of safe diagnostic tools and representative base sequences is available, the appropriateness of maintenance is actively debated because, in 1980, the World Heath Organisation declared smallpox (variola major) to be eradicated and there are concerns about the security implications.

Baculoviridae

This family of structurally and genetically sophisticated dsDNA (100–250 kbp) viruses has few, if any, relationships with known viruses of vertebrates. Baculoviruses have been recognized in arthropods including lepidopterans, dipterans, hymenopterans and crustaceans. The tendency for virions of baculoviruses to be embedded in a macroscopic proteinaceous matrix (polyhedron) is a feature which baculoviruses share with some other insect pathogenic viruses: entomopoxviruses and the cytoplasmic polyhedrosis reoviruses. Because baculoviruses continue to be named after the insect from which they were isolated, synonomy is undoubtedly prevalent. A few baculoviruses have been intensively studied (Ayres *et al.*, 1994), notably isolates which present opportunities for insect population management and/or have the capacity to produce recombinant proteins *in vitro* (King and Possee, 1992). The possible role of baculoviruses (genetically engineered or 'quasi-native') for insect management justified intensive investigation of environmental impact; one finding was the diversity of experimentally infectible hosts by one isolate (a nuclear polyhedrosis virus, NPV, from *Autographa californica*). Other baculoviruses are thought to have more limited host ranges (Blissard and Rohrmann, 1990) but the heterozygous potential provided by multiple virion envelopment coupled with multiple polyvirion occlusion in a polyhedron (which is characteristic of the baculovirus genus) has not been comprehensively investigated. One type of baculovirus is not naturally occluded (although, as a result of genetic engineering, this trait has been introduced from *A. californica* NPV; Crawford, 1989). Despite their non-occluded state, these baculoviruses have been arguably the most successful agent for (relatively unsophisticated) biological control of scarab beetles (notably *Oryctes rhinoceros* and *O. monosceros*) in coconut plantations.

Herpesviridae

The family takes its name from the virus causing recrudescent cold sores (herpes simplex) in humans; another predominantly human herpes virus causes chickenpox (varicella) in youth and shingles (zoster) in old age. However, herpes viruses have been identified from numerous non-human hosts including horses, rodents, canines, primates, snakes, frogs, fish, cattle and notably birds. Largely based on biological properties, Roizman *et al.* (1992) recognized three subfamilies (Alphaherpesvirinae, Betaherpesvirinae and Gammaherpesvirinae). In many instances animal cells have persistent infections in which virus particles are produced but herpes viruses are sometimes characterized by a nonproductive infection in which cells retain only the virus genome and survive. Immunosuppression to facilitate acceptance of genetically distinct tissue is a commonplace procedure in human medicine and, as a consequence, frank herpes-associated diseases are increasing worldwide. Blood products are screened routinely and are now a diminishing threat, but solid organ transplantation has substantial residual hazard from occult viruses: polyoma, parainfluenza and adenoviruses are all perilous whenever immunosuppression is used as a substitute for rigorous tissue matching. Bone marrow transplantation, requiring particularly profound immunosuppression, provides the greatest opportunities for recrudescence of endogenous herpes viruses.

Persistent herpes infections have been associated with tumours, e.g. Marek's disease in domestic fowl which is responsible for about 80% of virus-induced loss to the UK poultry industry. Duck plague is a herpes virus enteritis which killed about 42% of 100 000 wild mallard in Lake Andes, South Dakota in 1973 and 97% of ducks at a game farm in Wisconsin (Jacobsen *et al.*, 1976).

Iridoviridae

A 'family' of viruses with DNA genomes first recognized in association with a blue or green iridescence of invertebrate larvae. More recently, similar agents have also been recognized in fish (flounder lymphocystis), frogs, nematodes (Poinar *et al.*, 1980), crustaceans, isopods and possibly octopus. The 'family' is very diverse and too little studied to justify fixed classification, although iridoviruses with vertebrate hosts are distinguishable from one another and from a number of isolates from invertebrates (Williams, 1994). The agent of African swine fever which alternates between ticks and swine is now considered to be distinct from poxviruses and from iridoviruses, although there are superficial similarities between these three.

Papovaviridae

A family of viruses characterized by their propensity for causing tumours and named after papilloma virus causing self-limiting human warts and polyoma viruses causing tumours in mice. Inapparent infections seem normal. The role of human papilloma viruses in cervical cancers is incompletely understood: although there is a 90% association, there are, at present, no convincing data implicating papovaviruses, even with these human cancers (Turek, 1994). This is reassuring because, in the 1960s, a papovavirus (simian virus 40) was accidentally disseminated with contaminated polio vaccine; millions of people were inoculated.

Parvoviridae

Viruses assigned to this family and having single-stranded (ss)DNA genomes have been associated with fatal diseases of geese, dogs, pigs and lepidopterous insects (genus *Densovirus*). Horizontal (faecal–oral) but also vertical transmission of parvoviruses has been inferred and a diverse array inapparently infect rodents and felines; an additional range of parvoviruses is known only as contaminants of animal cells cultured *in vitro*. About 10 adeno-associated viruses (genus *Dependovirus*) which require co-infecting viruses in other families (Adenoviridae or Herpesviridae) to facilitate replication have been recognized in primates, cattle, dogs, horses and birds. Parvoviruses have been implicated with shellfish-associated food poisoning and porcine parvoviruses are a significant cause of abortion in pigs. However, the greatest perceived impact of parvoviruses has been in wild and commensal carnivores, notably dogs.

Geminiviruses

These viruses are characterized by a distinctive particle morphology (paired spheres or multiples up to four) and possession of circular ssDNA genomes. Some geminiviruses are transmitted between plants by whiteflies (Homoptera) whereas others, such as maize streak and sugarbeet curly top, have leafhopper (Homoptera) vectors. Before tolerant (but infectible) cultivars were widely grown, yield loss attributable to curly top in beets was in the range 4–10 tonnes/ha.

Reoviridae

The family name derives from a group of orphan agents with multisegmented dsRNA genomes. The viruses are widespread in human populations and readily isolated from people with respiratory or enteric diseases (but not proved to cause these conditions). Inapparent infection of dogs, birds, rodents and numerous other animals is commonplace. However, a few reoviruses cause respiratory illness in monkeys, and agents, grouped under the genus *Rotavirus*, incite diarrhoea in calves, mice and foals. It has been estimated that rotaviruses account for 1–2 million gastroenteritis-associated deaths in human infants of Asia, Africa and Latin America (Offit, 1994). Unfortunately, as with other infections of mucosal tissues, rotavirus infections induce only transient protection and prospects of effective protection by vaccination are not good.

An additional range of reoviruses replicating both in insects and in vertebrates, and assigned to the genus *Orbivirus*, are closely associated with diseases in humans and their domesticated animals. Two of the economically important reoviruses cause bluetongue of sheep and wild ruminants or African horse sickness. These viruses have dipterous insect vectors, but an additional range of orbiviruses is tick-borne. A notable feature of reoviruses is their predilection for invertebrates. The cytoplasmic polyhedrosis reoviruses have been recognized in about 150 different insect species and were associated with a silkworm disease of complex aetiology 'flacherie', which was estimated to have diminished Japan's silk yield by 31726 tons (6%) in 1960–1964. Additionally, a few (phyto)reoviruses replicate in both flowering plants (notably grasses) and insects, causing significant economic loss. In Italy, the planthopper-transmitted maize rough dwarf phytoreovirus was largely responsible for a loss estimated as $3 million at 1949 prices: during the 1960s about half of the hybrid maize grown in countries bordering the Mediterranean was infected, with some 20% diminishment in grain yield.

Togaviridae

The particles of viruses in this family are enveloped, hence the name (Latin *toga* = cloak), but in other respects they are a very mixed bag. Of the four genera presently recognized, two (the alphaviruses and most flaviviruses, recently designated Flaviviridae) infect both vertebrates and arthropods, whereas the others have no known invertebrate host. Non-arbo-togaviruses include several widespread but incompletely characterized agents such as lactate dehydrogenase-elevating virus (LDV), which is almost ubiquitous in mice and compromises experiments by contaminating murine cell lines. The agent of 'blue ear disease' in pigs (porcine reproductive and respiratory syndrome) is similar to LDV and was first recorded in Europe during 1990 (Meulenberg *et al.*, 1993; Done and Paton, 1995). The collective economic effects attributable to togaviruses are massive. Hog cholera (genus *Pestivirus*) is thought to have been responsible for more pig deaths worldwide than any other factor. However, during the past half century there seems to have been natural selection for diminished virulence. Before eradication from the USA, mortality varied to 100% and the virus was estimated to be capable of costing the US economy $100 million annually. In humans, rubella virus (causing German measles) is typically associated with a transient rash. Unfortunately, there are occasional serious sequelae, including stillbirth and fetal malformation. In the USA alone, the cost of institutional care and special education facilities needed for children born during 1964 and 1965 with congenitally acquired rubella was calculated to be $920 million.

The mosquito-transmitted viruses associated with equine encephalitis (genus *Alphavirus*) affected 184000 horses causing up to 90% mortality in the USA in 1938. Human mortality attributable to alphaviruses is fortunately not on this scale, but morbidity (infection incidence) is in many instances considerable. Five million Africans were affected by a disease that acquired the vernacular name of o'nyongnyong and, during 1962–1964, Venezuelan equine encephalomyelitis virus infected more than 30 000 people and countless horses; several hundred people died. The 30 or more viruses assigned to the genus *Flavivirus* include some having mosquito vectors, e.g. yellow fever, dengue and a suite of agents such as Kyasanur Forest disease virus which is carried by ticks. Despite safe and efficacious vaccination, yellow fever virus is an ever-present threat and is perhaps the most notorious, being responsible for deaths both in the neotropics and also in the Old World; the epidemic in Ethiopia during 1961–1962 probably involved 200 000 people of whom 30 000 died.

Alpha-like plant viruses

Alphaviruses of vertebrates have analogues which infect plants. These viruses are morphologically diverse and have different modes of spread but they have genome features in common. Among the currently designated genera are:

1. Tobamoviruses, which spread epidemically (to cause up to 20% loss of marketable yield in horticultural crops) when healthy plants rub against infected specimens. Individual species of the genus *Tobamovirus* have been isolated from wild plants (including flowering plants such as *Plantago* spp. and *Nicotiana velutina* and the alga *Chara austrialis*) but naturally infect few others.
2. Furoviruses, which cause potato mop-top, soil-borne wheat mosaic, beet necrotic yellow vein and peanut clump diseases, were distinguished from tobamoviruses largely because they are transmitted by soil-inhabiting fungi (*Plasmodiophorales*) that parasitize angiosperms (Cooper and Asher, 1988; Adams, 1991). In some instances the effects of furoviruses on yields are considerable. During 1957, in the state of Kansas, USA, soil-borne wheat mosaic furovirus caused the value of the wheat crop to be diminished by $4 million.
3. Closteroviruses, which are aphid-transmitted and have the longest virions yet detected (citrus tristeza closterovirus is a worldwide citrus hazard notorious for having caused the death of seven million citrus trees in Brazil).
4. Tobraviruses, which have tubular virions and spread with the aid of nematode (trichodorid) vectors.
5. Potexviruses which usually spread without vector assistance.
6. Tymoviruses, which have beetle vectors.

In addition, there is a subset of viruses which are unified by having tripartite genomes (currently designated Bromoviridae) and including cucumoviruses, bromoviruses and ilarviruses. Cucumber mosaic cucumovirus, which is transmissible by more than 60 species of aphids, is probably the most common virus in temperate crop plants and is known to be capable of infecting 775 species representing 85 plant families.

Orthomyxoviridae

A virus family having members assigned to three main groups. Type A has been recognized as a natural cause of diseases in primates, birds, horses, canines and pigs, whereas Types B and C have been recovered only from humans. The most notable disease attributable to an ortho-myxovirus is influenza which contributed to the death of 20 million people during 1919. Most years the effects of influenza on human populations are less drastic. Nevertheless, on the basis of costs in 1975, protective vaccination of postal workers in the UK was estimated to have a net cost benefit of £203 per 100 employees in the postal division and £70 per 100 for those in telecommunications. In the USA during 1966–1971, it was estimated that the cost per case of influenza A was in the range of $33–48 and, during 1977–1978, influenza B associated with headache, chill, rhinitis and myalgia as well as diarrhoea was projected from a 4.2% sample to have cost $30 per head.

Paramyxoviridae

Members of the family have some physicochemical properties in common with orthomyxoviruses but differ in having particles that are larger and a genome which is not segmented. The economic effects of paramyxoviruses are collectively very great. Mumps and measles are almost inescapable infections of childhood which are liable to be fatal, especially, but not exclusively, when associated with malnutrition. However, the economic effects of paramyxoviruses are most obvious in domesticated animals and poultry. Rinderpest/peste des petites rumin-ants are among the most damaging of infections affecting bovines, ovines, pigs, camels, etc. in Africa and Asia. Distemper/hardpad kills numerous canines, and Newcastle disease virus kills countless birds: 4 million in East Anglia in 1970. Eradication of Newcastle disease from California during 1971–1974 cost $56 million and the total cost due to death and protective vaccination in Singapore was put at $34.5 million per annum.

Rhabdoviridae

This is a family of viruses with negative-sense RNA genomes and members which naturally infect plants, insects, fish or mammals. Some rhabdoviruses that infect plants also replicate in their homopteran or heteropteran vectors whereas other rhabdoviruses multiply in dipterous flies as well as vertebrates. One of the most feared rhabdovirus-associated diseases is rabies, which is estimated to kill about 15 000 people per year worldwide. Whereas the number of human deaths in Europe is fewer than 10 per year, the rabies problem is on a much grander scale in the neotropics and India. In Central America, where approximately 250 000 people are treated each year because of possible infection, the direct economic loss reportedly exceeds $250 million annually and indirect costs attributable to malnutrition through livestock loss has been judged to be 1000-fold greater. Although most vertebrates are infectible by rabies virus, urban humans are most at risk when dogs are infected, and strenuous efforts are devoted to local eradication of sources of infection. The prophylactic and containment costs attributable to one rabid dog in California during 1980 were estimated as $92 650 for human treatment, $4190 for animal vaccination and $8950 for dog-catching and destruction. In comparison to this, rhabdovirus-induced plant diseases are minor inconveniences. Nevertheless, the piesmid-transmitted sugar beet leaf curl virus was estimated to have caused losses of $12 million in Poland during 1957.

Bunyaviridae

These are a family of viruses which multiply in arthropods (e.g. mosquitoes, sandflies, midges and ticks), in some instances implicated as vectors to humans or other animals (Bishop and Shope, 1979; Griott et al., 1993). Epidemics of Rift valley fever and Nairobi sheep disease attributed to bunyaviruses have great economic importance but, with these viruses, human infection only occasionally manifests itself as illness or death. Tomato spotted wilt virus includes a suite of isolates (genus *Tospovirus*) which are transmissible between plants by thrips (*Thysanoptera*). The experimental host range of tospovirus isolates embraces some 500 species in 70 plant families, making management very difficult (German et al., 1992). As a consequence of dispersal in trade, of an unusually effective vector (*Frankiniella occidentalis*), tospoviruses have become pandemic and almost justify description as 'a biblical plague'. Tospoviruses have genome properties (tripartite ssRNA with negative or ambisense polarities) resembling vertebrate bunyaviruses but contain an extra gene which facilitates their systemic invasion in plants.

Retroviridae

The retroviruses are characterized by their possession of an enzyme (reverse transcriptase) that transcribes virus-associated RNA into an infectious DNA species which may be incorporated into the DNA of host cells. Four categories of virus use this tactic; retroviruses, and the pararetroviruses which are distinguished by encapsidating DNA and include hepadnaviruses of vertebrates and two groups of non-enveloped viruses (caulimoviruses and badnaviruses) which infect plants (Rothine *et al.*, 1994). The retroviruses have several members associated with cancers of birds or mammals and a few that have been isolated from reptiles or fish. Virus-like particles in thin sections of infected cells examined by electron microscopy were divided into four types A, B, C and D. There is a substantial body of evidence associating type C retroviruses with human leukaemias and, more circumstantially, type B particles with breast cancer. Simian acquired immune deficiency (SAIDS) is caused by a type D retrovirus and differs from isolates of the Lentivirus subfamily of the Retroviridae which are characterized by morphologically distinctive virions (type C). Lentiviruses have major importance because they cause chronically debilitating illness (acquired immune deficiencies; AIDS) in a variety of animals: felines, rodents, ungulates and primates. Human immunodeficiency viruses were disseminated catastrophically in blood products. Analogously, but less immediately life-threatening, the agent of type B hepatitis debilitated 45 000 US military personnel following administration to them of contaminated yellow fever vaccine. Zuckerman and Howard (1979) estimated that hepatitis B infected about 200 million people: infection is associated with the most common fatal human cancer, primary hepatic carcinoma (see also Ellis, 1993).

Picornaviridae

The family name for these viruses reflects their small size (Greek *pico* = small) and constituent RNA. Dozens of picornaviruses assigned to four main genera have been recognized worldwide in association with transient illness of high morbidity, e.g. the numerous common cold viruses (rhinoviruses) and hepatitis A have been assigned to the genus *Enterovirus* in the Picornaviridae. Inapparent infection is probably very frequent in a diverse range of vertebrates (e.g. cardioviruses naturally infecting rodents, swine and occasionally captive primates). However, the Picornaviridae includes viruses with a marked tendency to cause debilitating or fatal illness. Some viruses assigned to the *Enterovirus* genus cause neurological disorders including poliomyelitis. Although the suffering associated with protracted incapacity of children cannot be quantified, it has been estimated that the loss of current and potential earnings resulting from poliomyelitis during 1957–1958 amounted to £4.3 million in the UK. Additionally, picornaviruses significantly lessen food production by cloven-footed animals, even though the diseases they cause are not usually themselves fatal. The enterovirus causing swine vesicular disease and the suite of agents (in the genus *Aphthovirus*) associated with foot-and-mouth disease are not always distinguished but it is noteworthy that the direct cost of the 1967–1968 epidemic of foot-and-mouth disease in the UK was estimated at £35.1 million, largely attributable to the cost of compensation for slaughter (to control virus spread) but also valuation, cremation/burial and disinfection. The indirect costs attributable to loss of income from the slaughtered animals and disruption of agricultural production, marketing and distribution provide scope for wide divergence of view but were in the region of £100 million.

The Picornaviridae is a family that has been established longer than many others. Predictably, more recent research showed that some of the viruses have substantially distinctive properties. Among the outgroups ICTV recently recognized two additional families: Nodaviridae (having invertebrate hosts and bipartite RNA genomes) and Caliciviridae.

Caliciviridae

Caliciviruses include isolates from cats, rabbits, swine, mink and marine mammals such as seals. Until recently, the most significant disease associated with caliciviruses was vesicular exanthema in swine which, during 1952–1955, was estimated to have cost the USA $33 million. Circumstantial evidence has linked infected pork with abortion in sealions and pinniped carcasses with the primary porcine infection. During the past decade a hitherto uncommon and localized calicivirus responsible for a major pandemic of fatal haemorrhagic disease has spread (possibly in frozen rabbit meat for cat food) from a focus in China. The associated virus is being considered as an alternative to myxoma poxvirus for the management of wild rabbit populations in Australia.

Picorna-like plant viruses

The picorna-like viruses of plants were among the first to be distinguished from the picornaviruses of vertebrates but, despite similarities in genome organization, substantial diversity is now being revealed and revision of the current nomenclature is likely to be necessary. Vector differences and virion morphology were the bases for segregating the Potyviridae from the picorna-like plant-infecting viruses with bipartite genomes (Comoviridae). However, as more sequence data become available, the designation of genera (*Fabavirus, Nepovirus* and *Comovirus*) based on vector relations (aphids, nematodes and beetles respectively) is being seen as premature. The Potyviridae, as presently recognized, has three ICTV-approved genera distinguished largely on the basis of vector. Virus species in the genus *Potyvirus* are transmitted transiently by aphids, bymoviruses are transmitted between grasses by plasmo-diophoromycete fungi and rymoviruses have eriophyid mites as vectors. An additional group of potyviruses unified by having whitefly vectors has been proposed (Shukla *et al.*, 1994) and tentatively designated as the genus *Ipomovirus*. Viruses in the Comoviridae are prevalent in wild and cultivated plants but not dramatically damaging. By contrast, among the 200 potyviruses currently recognized there are about 80 species that collectively include most plant species and families as hosts. Plum pox and potato viruses A and Y are all economically damaging potyviruses that spread rapidly and, depending on the cultivar affected (and the number and sort of co-infecting viruses) diminish marketable yield by 80–99%. In California, lettuce mosaic potyvirus causes losses estimated at $1000/ha directly plus an equal or greater amount due to under-/over-production, fluctuating availability and price. Damage to cucurbit crops by potyviruses (notably zuccini yellow mosaic and watermelon mosaic 2) is on a similar scale. Whereas potyviruses generally have monopartite ssRNA genomes, bymoviruses, which can be locally devastating, are bipartite: their genomes comprise a larger potyvirus-derived segment (RNA-1) as well as a smaller (RNA-2) with a distinctive organization (resembling that of furoviruses or tobamoviruses) and coding for fungus association (Dessens *et al.*, 1995).

As the base sequences in an increasing number of viral genomes have been elucidated, rationalization of the diversity is being attempted. Analyses of genome sequences revealed evidence consistent with recombination which seems to be particularly promiscuous among the *Tombusvirus* and *Carmovirus* genera of the Tombusviridae, sobemoviruses and umbraviruses (Gibbs and Cooper, 1995). Luteoviruses, which have lineages that embrace a variety of extant taxa and for which current virus nomenclature is particularly inadequate, are some of the world's most economically damaging plant pathogens. The viruses are transmitted by aphids (*Homoptera*) in which they do not replicate but are retained at least partly as a result of interactions with products of endosymbiotic prokaryotes in the vectors (van den Heuvel *et al.*, 1994). In the Pacific north-western states of the USA during the period 1951–1960, the average annual loss due to barley yellow dwarf luteovirus (*sensu lato*) was $6 million in barley and $36 million in oats. In the same decade, luteoviruses associated with western yellows/mild yellowing in sugar beets were judged to cause losses of $10 million and potato leaf roll caused an additional loss of $14 million annually.

1.6 CONSEQUENCES OF MULTIPLE INFECTION

1.6.1 POLYMORPHISM

A single cell may be simultaneously infected experimentally by two distinct viruses and in nature, organisms are commonly infected by two or more viruses (Abdalla *et al.*, 1985) which may interact imperceptibly, antagonistically (Huang and Baltimore, 1977; Holland *et al.*, 1980) or synergistically. Before the advent of virus-derived transgenic genes in crop plants, opportunities for viral interactions depended on multiple infections which were constrained by vector and host preferences. Now transgenic plants potentially provide more fruitful fields for evolution.

Virus-derived genes in plants tend to lessen disease severity (i.e. tolerance *sensu*; Cooper and Jones, 1983) but there is not always a commensurate lessening of virus replication/accumulation. Furthermore, any 'protection' associated with the transgenic gene tends to be virus-specific (although there are a few exceptions, e.g. Dinant *et al.*, 1993). Because plants with transgenic virus-derived genes are infectible by diverse viruses, concerns have been expressed about the implications. Since capsid coding sequences from viruses are conveniently manipulated, they are frequently used and plants containing such genes are likely to be among the first to be commercialized. One consequence of virus infection in such capsid expressors is resurfacing: the incorporation of one virus genome into protein coats or lipid envelopes synthesized wholly or partly in response to the presence of a second virus genome. Resurfacing has been recognized as a natural outcome of multiple infection in diverse systems: during growth of bacteriophages, enteroviruses, adenoviruses, orthomyxoviruses, paramyxoviruses, alphaviruses and among luteoviruses (Creamer and Falk, 1990; Wen and Lister, 1991) or potyviruses (Bourdin and Lecoq, 1991). There are also examples of transcapsidation with products of transgenic genes (Candelier-Harvey and Hull, 1993) in plants. Transcapsidation is potentially important because of the role played by capsid proteins in vector attachment at least for some viruses (e.g. geminiviruses; Briddon *et al.*, 1990) but also because capsids protect vulnerable nucleic acids and are in some instances essential facilitators of systemic invasion by viral nucleic acids (Wellink and van Kammen, 1989). However, even when transcapsidation occurs in plants containing virus-derived transgenic genes, the outcome need not be different from that which accompanies 'normal' virus replication (Cooper *et al.*, 1994). Figure 1.3 indicates generalized names given to some products of double infection. When infected cells contain different viruses, the extent to which resurfacing occurs varies but is probably greatest when the individual virus geno-types have compatible replication requirements and where the virus particles are structurally similar (Dodds and Hamilton, 1976).

There are no unequivocal data on the proportions of particles that are phenotypically mixed or genomically masked during multiple infec-tions; it is technically very difficult to quantify the numbers of cells in a multicellular host which are infected singly. The relative abundance of the nucleic acid and structural components of viruses replicating together is another determining factor which is difficult to quantify. Even within a cell there may be scope for compartmentalization which constrains opportunities for interaction. The available data are few and widely different but such gene capture almost certainly facilitates con-tinued association between virus nucleic acids with distinct evolutionary backgrounds and will thereby tend to facilitate recombination and evolution. Thus, at two extremes, the percentage of non-matching genotypes for a given phenotype (genomically masked particles) vary from 0.00001% for foot-and-mouth disease virus to 96% for the 'satellite' of tobacco ringspot virus. Satellitism reflects dependence on a co-infect-ing agent and is a phenomenon exhibited by a diverse range of viruses, including some that are plant pathogens (Murant and Mayo, 1982; Liu and Cooper, 1994) and others that are associated with vertebrate pathogens such as the adeno-associated parvoviruses (Berns and Hauswirth, 1979). Because surface properties of a virion determine attachment and hence whether or not infection occurs, resurfacing may lead to altered host range. This seems to have occurred when poliovirus RNA was coated *in vivo* with protein from another enteric picornavirus and some of the viral products infected mice, which are unnatural

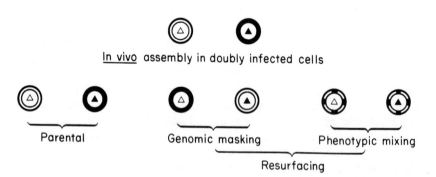

Figure 1.3 Names given to products of *in vivo* assembly of viruses in doubly infected cells.

hosts for poliovirus. Similarly, avian leukoviruses, such as Rous sarcoma, lacking surface (envelope) properties facilitating infection of avian cells have been observed to infect birds when genomically masked with the envelope coded in a distinct isolate of the same virus. Additionally, resurfacing facilitates the carriage of heterologous nucleic acids. Packing specificities impose constraints in many viruses but a phage has been reported to incorporate 10% of the genome of its bacterial host, a phenomenon called transduction in this context. Resurfacing of DNA genomes with proteins coded by co-infecting RNA viruses (or vice versa) is not uncommon (Zavada, 1982); indeed, some capsid proteins seem to preferentially encapsidate host-derived nucleic acid species (Rochon et al., 1986).

1.6.2 INTERFERENCE

In vertebrates, there are several types of interference between co-infecting viruses. Some result in the production of interferon and others result from incompatible replication requirements, as when poliomyelitis virus inactivates cap-binding which is needed by vesicular stomatitis rhabdovirus (VSV) or when VSV shuts down host transcription thereby depriving influenza A of primers which are involved in RNA synthesis by this virus. A more sophisticated exclusion system operates with orthomyxoviruses and paramyxoviruses. They encode a repressor destroying protein that facilitates egress of their progeny while tending to prevent binding and hence infection by other viruses with the same portal of entry.

In plants, co-infections seem commonplace in nature but have usually been noticed because of abnormal pathology in short-lived herbaceous plants. Although co-infections do not invariably correlate with virulence, interactions between viruses in plants have revealed impact on virus concentration and distribution – both features with potential epidemiological significance. Co-infections with 'similar' viruses are thought to be uncommon, although discriminating and sensitive tests which would select such mixed infections have been used too infrequently to provide a reliable basis for judgement.

In vertebrates, the rarity of re-infection by related viruses is usually attributed to humoral and cellular factors which circulate following primary exposure/infection. However, by no means are all viruses equally good at eliciting such responses and many elicit only transient protection. In a few crops (e.g. citrus; Bar-Joseph, 1978), a similar phenomenon ('cross-protection') is routinely exploited by inoculation (resembling active immunization) with apathogenic viruses (see Chapter 6). The basis for cross-protection in plants remains unknown but there are parallels between the protection which it affords and that due to transgenic 'pathogen-derived' resistances (*sensu* Sanford and Johnston, 1985).

1.6.3 QUASI-GENETIC INTERACTIONS RESULTING FROM HETEROZYGOSITY

The phenomenon of resurfacing has been studied in greatest depth and the implications seem greatest for viruses with RNA genomes. One consequence of resurfacing may have been the stabilization and maintenance of extraneous nucleic acids or subgenomic fragments that formed by chance. However, whether or not this explains how divided genomes evolved, it is known that some plant pathogens with divided single-stranded, positive-sense RNA genomes exhibit genetic linkage (see Figure 1.4). Re-assortants have adaptive significance and heterozygosity is likely among species of bunyaviruses, orthomyxoviruses, arenaviruses and reoviruses. Retroviruses are functionally diploid but in essence most other viruses are haploid or exceptionally polyploid, as in baculoviruses. When genes facilitating different options for dispersal are detached from one another, heterozygosity potentially facilitates re-assortment, enhances evolution and is a valuable survival strategy.

There is little evidence that recombination by breakage and reunion of nucleic acid occurs in viruses containing single-stranded RNA. High mutation rates coupled with extra fecundity distinguish RNA genetics from DNA genetics but there are structural and population limits on the abundance of mutation (Holland *et al.*, 1992; Clarke *et al.*, 1993; Morse, 1994). Much primary genetic variation is attributable to single base changes which are much more common during RNA than DNA replication. However, breakage leading to deletion is more noticeable – particularly in some viruses (e.g. fungus-transmitted furoviruses) in which deletion variants rapidly accumulate during infections. This sort of genetic change also can be detected very frequently in tobacco necrosis necrovirus (TNV) and produces mutants which cannot synthesize their own capsid protein. Consequent upon genomic masking, coatless mutants produced in this way are maintained as a result of co-infection with 'normal' TNV (capable of coding for coat protein). Luteoviruses have an analogous division of labour with their umbraviral coviruses. Whereas luteoviruses code for capsid proteins which are variable (Bahner *et al.*, 1990) but essential for transmission by aphids, they do not contain a gene that facilitates cell–cell invasion. By contrast, umbraviruses do not code for a capsid protein but they do contain a gene which facilitates cell–cell invasion by the umbravirus and also the escape of co-infecting luteoviruses from the vascular tissues that were inoculated by virus-carrying aphids. Possibly because extra cells become available for virus replication, the presence of umbraviruses has been

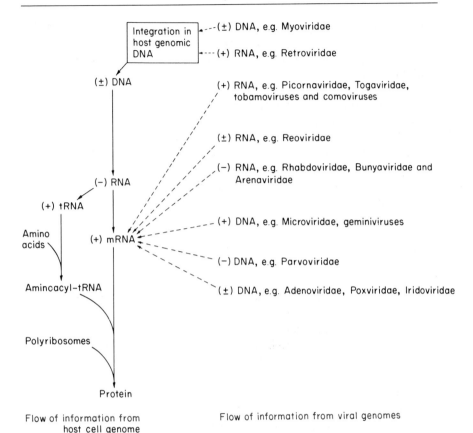

Flow of information from   Flow of information from viral genomes
host cell genome

Figure 1.4 Diagram indicating routes via which information encoded in the sequence of purine/pyrimidine bases in viral genomes is expressed as functional proteins. By convention the base sequence of messenger (m)RNA which is translated into a characteristic succession of amino acids is described as positive sense (+), and its complement in which adenine is paired with uridine in double-stranded (±) RNA or with thymidine in double-stranded (±) DNA, is negative sense (−).

correlated with a ten-fold increase in the accumulation of luteoviruses (Barker, 1989) which thereby have enhanced prospects for aphid association and subsequent dispersal to other plants. The mutualistic interaction between the genomes facilitates their coevolution and such a process can be seen as a forerunner of multipartite genomes that have been adopted by numerous viruses. Analogous heterozygosity occurs in tobravirus populations in which the two genomic RNAs are packaged separately in tubular virions with lengths reflecting the encapsidated RNA and as more viral genome sequences are determined, the prevalence of the phenomenon is likely to become more apparent. Like the umbraviral genome, tobravirus RNA 1 is independently infectious for plants but can be equated to a deletion mutant because it does not code for a capsid protein. It is difficult to judge the importance of these RNA-based agents, but, at least in perennial hosts, they potentially represent pre-adapted genomes which can re-enter viral gene pools when their repository is coinfected with a compatible virus.

The *in vitro* combination of RNA 1 from one tobravirus isolate with RNA 2 from another tends to result in new infectious associations called pseudorecombinants. From knowledge of the properties possessed by the progenitors and the pseudorecombinants, a variety of genetic markers have been located on one or other of the two sub-genomic fragments (Harrison and Robinson, 1988).

Nepoviruses (Comoviridae) resemble tobraviruses in having bipartite genomes in which the genetic determinants for coat protein are located on the smaller RNA 2. With nepoviruses, seed transmissibility in chickweed (*Stellaria media*) is a property largely determined by RNA 1. Consequently, two options for virus dissemination (via seed or nematodes) are controlled separately. Predictably, nepovirus genome combinations detected in plants that were infected naturally differ from laboratory-synthesized pseudorecombinants in possessing a range of properties that facilitate their dispersal and survival. Presumably, any novel feature a viral genome may acquire by chance association takes time to accumulate the appropriate genetic background with which it can integrate effectively. The pace of this acquisition of balanced geno-type seems to be slow. More than 20 years ago, a nepovirus isolate with a hitherto unrecognized trait (the ability to infect the raspberry cultivar, Lloyd George) was first noticed. Although the isolate was subsequently shown to be transmitted by nematodes (*Longidorus elon-gatus*), so far as is known, this virus has remained restricted to a very circumscribed region of longidorid-infested soil within an area of conti-guous raspberry production. Properties of the virus that might explain this restriction include infrequent seed transmissibility (Hanada and Harrison, 1977).

Enhanced levels of genome complexity are known (e.g, furoviruses, bromoviruses, cucumoviruses and ilarviruses) and experience gained with these viruses indicates that viruses *in vivo* are by no means genetically fixed and populations are fluid to an extent that is impossible to measure reproducibly.

Appendix 1.1 Characteristics of some virus families and groups described in the text

Genome type/number of parts	Particle morphology (+/−) envelope	Family/group	V*	B*	P*	I*	F*	A*	M*
ssRNA (1)	isometric (−)	Picornaviridae	+	−	+	+?	−	−	−
ssRNA (1)	isometric (−)	Caliciviridae	+	−	−	−	−	−	−
ssRNA (1)	isometric (−)	Leviviridae	−	+	−	−	−	−	−
ssRNA (1)	isometric (−)	Tetraviridae	−	−	−	+	−	−	−
ssRNA (1)	helical (−)	Potyviridae	−	−	+	−	−	−	−
ssRNA (1)	helical (−)	Tobamovirus	−	−	+	−	−	+?	−
ssRNA (1)	helical (−)	Potexvirus	−	−	+	−	−	−	−
ssRNA (1)	helical (−)	Carlavirus	−	−	+	−	−	−	−
ssRNA (1)	helical (−)	Closterovirus	−	−	+	−	−	−	−
ssRNA (1)	helical (−)	Capillovirus	−	−	+	−	−	−	−
ssRNA (1)	isometric (−)	Tymovirus	−	−	+	−	−	−	−
ssRNA (1)	isometric (−)	Tombusvirus	−	−	+	−	−	−	−
ssRNA (1)	isometric (−)	Sobemovirus	−	−	+	−	−	−	−
ssRNA (1)	isometric (−)	Carmovirus	−	−	+	−	−	−	−
ssRNA (1)	isometric (−)	Necrovirus	−	−	+	−	−	−	−
ssRNA (1)	isometric (−)	Luteovirus	−	−	+	−	−	−	−
ssRNA (2)	isometric (−)	Comoviridae	−	−	+	−	−	−	−
ssRNA (2)	isometric (−)	Dianthovirus	−	−	+	−	−	−	−
ssRNA (2)	helical (−)	Furovirus	−	−	+	−	−	−	−
ssRNA (2)	helical (−)	Tobravirus	−	−	+	−	−	−	−
ssRNA (3)	isometric (−)	Bromovirus	−	−	+	−	−	−	−
ssRNA (3)	isometric (−)	Ilarvirus	−	−	+	−	−	−	−
ssRNA (3)	isometric (−)	Cucumovirus	−	−	+	−	−	−	−

Appendix 1.1 *Continued*

Genome type/number of parts	Particle morphology (+/−) envelope	Family/group	V*	B*	P*	I*	F*	A*	M*
ssRNA (3)	helical (−)	Hordeivirus	−	−	+	−	−	−	−
ssRNA (4?)	helical (−)	Tenuivirus	−	−	+	−	−	−	−
ssRNA (2)	isometric (−)	Nodaviridae	+	−	+	+	−	−	−
ssRNA (1)	pleomorphic (+)	Coronaviridae	+	−	−	−	−	−	−
ssRNA (1)	isometric (+)	Flaviviridae	+	−	−	+	−	−	−
ssRNA (1)	isometric (+)	Togaviridae	+	−	−	+	−	−	−
ssRNA (1)	bacilliform (+)	Filoviridae	+	−	−	−	−	−	−
ssRNA (1)	bacilliform (+)	Rhabdoviridae	+	−	+	+	−	−	−
ssRNA (1)	helical (+)	Paramyxoviridae	+	−	−	−	−	−	−
ssRNA (2)	isometric (+)	Arenaviridae	+	−	−	−	−	−	−
ssRNA (3)	isometric (+)	Bunyaviridae	+	−	+	+	−	−	−
ssRNA (8)	helical (+)	Orthomyxoviridae	+	−	−	−	−	−	−
ssRNA (1+)	isometric (+)	Retroviridae	+	−	−	−	−	−	−
dsDNA (1)	bacilliform (+)	Baculoviridae	−	−	−	+	−	−	−
dsDNA (1)	isometric (+)	Hepadnaviridae	+	−	−	−	−	−	−
dsDNA (1)	isometric (+)	Herpesviridae	+	−	−	−	−	−	−
dsDNA (1)	cubicoid (+)	Poxviridae	+	−	−	+	−	−	−
dsDNA (1)	helical (+)	Polydnaviridae	−	−	+	+	−	−	−
dsDNA (1)	phage with tail (−)	Myoviridae	−	+	−	−	−	−	−

Genome	Morphology	Family	V	B	P	I	F	A	M
dsDNA (1)	phage with tail (−)	Podoviridae	−	+	−	−	−	−	−
dsDNA (1)	phage with tail (−)	Siphoviridae	−	+	−	−	−	−	−
dsDNA (1)	isometric (−)	Adenoviridae	+	−	−	−	−	−	−
dsDNA (1)	isometric (−)	Caulimovirus	−	−	+	−	−	−	−
dsDNA (1)	bacilliform (−)	Badnavirus	−	−	+	−	−	−	−
dsDNA (1)	isometric (−)	Iridoviridae	+	−	−	+	−	−	−
dsDNA (1)	isometric (−)	Papovaviridae	+	−	−	−	−	−	−
dsDNA (1)	isometric (−)	Phycodnaviridae	−	−	−	−	−	+	−
dsDNA (1)	isometric (−)	Tectiviridae	−	+	−	−	−	−	−
ssDNA (1)	helical (−)	Inoviridae	−	+	−	−	−	−	+
ssDNA (1)	isometric (−)	Microviridae	−	+	−	−	−	−	−
ssDNA (1)	isometric (−)	Parvoviridae	+	−	−	+	−	−	−
ssDNA (1/2)	isometric (−)	Geminivirus	−	−	+	−	−	−	−
dsRNA (1)	isometric (−)	Totiviridae	−	−	−	−	+	−	−
dsRNA (2)	isometric (−)	Birnaviridae	+	−	−	+	−	−	−
dsRNA (2)	isometric (−)	Cryptoviridae	−	−	+	−	−	−	−
dsRNA (2)	isometric (−)	Partitiviridae	−	−	−	−	+	−	−
dsRNA (3)	isometric (+)	Cystoviridae	−	+	−	−	−	−	−
dsRNA (10/12)	isometric (−)	Reoviridae	+	−	+	+	−	−	−

*Key:-V, vertebrate host; B, bacterial host; P, plant host; I, invertebrate host; F, fungal host; A, algal host; M, mollicute host

Exposure to viruses and some consequences

2

2.1 INOCULATION

The act of introducing a potentially infectious agent into an organism is usually described as inoculation. The word inoculation derives from the insertion of an eye (or bud) into a plant and recognizes an ancient horticultural practice that commonly results in the transmission of disease-causing agents. In nature, tissue fusion facilitating virus trans-mission happens routinely in some fungi (Basidiomycetes and Asco-mycetes). Analogously, grafting of tree roots is surprisingly common especially in stony soil but despite the opportunities, consequential inoculation of trees by viruses seems infrequent (Cooper, 1993). Fur-thermore, angiosperm parasites such as dodders (*Cuscuta* spp.) which graft themselves on to other plants are known to be able to inoculate simultaneously a range of viruses, including several which do not replicate in the dodder en route (Bennett, 1967a).

Grafting also facilitates inoculation of vertebrates. Indeed, no surgical procedure is without risk; ophthalmological investigations can effect inoculation at the conjunctiva and all instruments (including acupuncture/tattooing needles and earrings) inserted into skin are exposed to contamination with viruses which occur in blood before signs of disease are apparent. The routine use of organ graft or trans-plantation of tissues, including blood vessels, cornea, kidney, liver and heart, from donors that died only hours before has enhanced the possibility of accidental transmission. Fortunately, when the surgery is not an emergency and storable tissues (such as blood vessels or eyes) are used, tests on the donor are possible and the risk is minimal. Nevertheless, one instance of human inoculation with rabies virus is plausibly attributed to such grafting. Polyoma and adenoviruses are not uncommonly transmitted by grafting and leucoviruses con-taminating vaccines based on egg-adapted isolates also carry a small

risk. However, in the context of normal grafting, inapparent herpesviruses present the greatest peril. Fortunately, rapid laboratory
detection methods are available and while the danger to recipients is
small, it is inevitably magnified because of the concomitant treatment
to prevent tissue rejection. Immunosuppression to facilitate acceptance
of genetically distinct tissue is increasingly practised and favours
recrudescence of endogenous (latent) viruses. Transplantation of solid
organs has a 20% case prevalence of recrudescent cytomegaloher-
pesvirus. Even though 90% of adults are seropositive for Epstein–Barr
herpesvirus, the agents of hepatitis B and hepatitis C present more
notable hazards in blood/liver transplantation, particularly in south-
east Asia (e.g. Taiwan) where prevalence is an order of magnitude
greater than in, for example, the USA.

To alleviate the shortage of human organs for transplantation and,
following genetic engineering, use of 'humanized' non-human organs
(e.g. from pigs) may become routine – notwithstanding the ethical
concerns. There may be virological hazards associated with this technology but these are modest by comparison with those which might
result from the use of primate tissues; baboon bone marrow is being
considered for the repair of damaged human immune systems. Clearly,
having regard to the known prevalence in primates of viruses pathogenic for humans, exhaustively detailed and long-term safety tests will
be essential before this opportunity is followed up.

2.2 INFECTION

When an organism is inoculated with material containing viruses there
are two possible consequences: either infection occurs or it does not.
Nucleoprotein virus particles can enter cells in which they do not
replicate and their constituents may move separately or collectively
within cells entering nuclei, chloroplasts, etc. However, these acts do
not constitute infection, which is a term properly restricted to the act of
nucleic acid replication following nucleic acid or nucleoprotein entry
into cells. There is no unanimously agreed term to describe either the
absolute state of exemption from infection or the antithesis of the state.
To distinguish host from non-hosts unambiguously Cooper and Jones
(1983) recommended the term **infectible**. **Immune** would be an ideal
antonym, but the word and its derivatives have long been used un-
critically to denote freedom from disease. Additionally, **immunity**,
immunology, etc. have well-established usage in the description of
blood-associated factors that facilitate defence against a diverse range
of infectious agents. The behaviour of a virus in a host is often confused
with the responses of the host-to-virus infection. We tried to reserve
the terms **resistant** and **susceptible** to denote opposite ends of a scale

covering the effects of an infectible individual on virus infection, multiplication and invasion. The terms **tolerant** and **sensitive** were recommended to denote the opposite ends of a scale covering the disease reaction to virus infection and establishment. Several distinguishable components of resistance have now been identified. Thus, passive resistance to infection by virus occurs when an infectible individual is not very readily infected. For example, the incompatibility of surface properties possessed by virus nucleoprotein for surfaces of potentially infectible cells may hinder adsorption or attachment. In reality, plants are not known to have cell surface receptors facilitating virus attachment. However, the role may be occupied by the sites inside cells at which virions accumulate before migrating from one cell to neighbours. These putative sites are membrane-associated and usually equated with virus-coded 'movement' proteins (of which there are several distinct species). Once virus nucleic acid has entered a cell, the availability of enzymes facilitating translation/replication may impose constraints either on the occurrence or rate of virus multiplication. Because the activity (and perhaps the occurrence) of such enzymes is responsive to environmental conditions such as temperature, cells that are infectible in some circumstances may not be infectible in others.

Genetic diversity facilitates adaptation of potential hosts to evolutionary pressures in which it is presumed that the species, rather than individuals, benefit. Experiments involving fecund vertebrates such as mice have helped to identify polymorphisms which impart resistance to challenge by a virus genotype. The best studied traits regulate infectibility by influenza A and, on interferon induction, directly inhibit replication of this virus (Reeves *et al.*, 1988). Plausibly, the analogous properties in plants equate with major genes for 'resistance' (without invoking interferon). In bacteria, there is a variety of host restriction systems that either prevent primary infection, establishment, progeny assembly or superinfection. Predictably, there is a struggle for gene supremacy and viruses tend to perpetuate genomes (with modified bases) which circumvent host attack by nucleic acid-modifying enzymes that are particularly commonplace in prokaryotes, although much less so in eukaryotes. Phages remember (by perpetuation after selection) the genotypes of hosts in which they reproduced. The biology of host restriction and the enzymes involved (which revolutionized molecular biology) have been discussed by Bickle and Kruger (1993).

Information concerning virus infection is scarce but there seem to be common features whether virions have DNA or RNA genomes; both virus-coded and host-encoded functions are involved. The first cell inoculated is only a temporary reservoir for viruses that resume their replication when they have crossed interfaces with adjoining cells. It has been claimed that rod-shaped plant pathogenic viruses adsorb end-on to leaf surfaces (Bukrinskaya, 1982). Whether this is common or not, plants (apart perhaps from planktonic unicellular motile organisms such as fungal zoospores to which virus-like particles attributable to tobacco necrosis necrovirus attach at flagellae) differ from bacteria and animal cells in lacking specific recognition sites on their surfaces to which virus particles adsorb. Plants cells are routinely coated by cutins, waxes and a rigid cellulose cell wall. Following abrasion, these barriers can be penetrated by viruses. However, the viruses do not spread except when they trigger the expansion of gates across tubules which link adjoining cells (Citovsky and Zambyski, 1993). The spreading trait is usually assumed to be virus-coded and represents a feature which distinguishes viruses 'of' plants from structurally and compositionally similar viruses 'of' vertebrates.

Animal cells lack the structural constraint provided by a rigid cellulose wall and have the capacity for endocytotic engulfment. Even though endocytosis potentially delivers infectious virions into the cytoplasm, the balance of evidence suggests that it is nucleic acid liberated from such foci that initiates the replication process. A variety of receptors has been recognized as facilitators of bacteriophage attachment to bacterial surfaces or the attachment of vertebrate viruses to specific cells. Cell surface receptor molecules for human immunodeficiency lentiviruses, rabies, vaccinia and some reoviruses have been identified and include immunoglobulins or cell growth modulators. Such observations cast light on mechanisms of immunosuppression on the one hand and disorganized growth (cancer transformation) on the other (Sauve et al., 1993; Geleziunas et al., 1994).

In plants, virions circulate within the water and particularly the food conduction tissues (phloem). Furthermore, the presence of virus-like particles within transwall tubules and the evidence from mutational analysis of potyviruses suggests the capsid protein might have a crucial role related to cell–cell movement. Knowledge about the mechanisms whereby virions push or pull their way into their next cell is very slight. A variety of mechanisms and virus-coded genes have been invoked for differing viruses but in many instances only the nucleic acid penetrates cells of a potential host. Active injection of genomic nucleic acids, as with some viruses of bacteria (cf. Hershey and Chase, 1952) is a highly evolved, specialized and seemingly exceptional process. In viruses with single-stranded RNA genomes, it is not known whether the translatable (positive-sense) or the complementary strand generated during replication is the actual form in which a virus (possibly in combination with protein) invades an adjoining cell. Nevertheless, since virion-derived positive-sense, single-stranded RNA is a reliably infectious experimental inoculum, it would not be surprising if nucleic acids only partially disassociated from their capsid proteins are biologically active in this context as has been suggested by Wilson and co-workers (Wilson, 1984). Some evidence suggests that nucleic acids are released by a rapid uncoating (physical) process which is 25% complete within 10 minutes (for tobacco mosaic tobamovirus). In plants containing capsid protein-deficient viruses (e.g. umbraviruses), viroids or subgenomic (RNA-1) forms of tobraviruses, nucleic acid is almost certainly the entity that infects cells and, with varying degrees of efficiency, systemically invades tissues. It is by no means certain that the virions themselves are very important except as vehicles that stabilize and protect viral genomes en route between potential hosts, a journey that is often effected by another organism – a vector.

2.3 INVASION

The primary acts that result in the establishment of a parasite–host relationship are distinguishable from subsequent events that lead to invasion of other cells and which may or may not lead to overt disease (Figure 2.1). Most virus infections are more or less localized and inapparent depending on the genetic constitution of the host, age and numerous other factors including the virus concerned.

Overt disease or cell dysfunction that facilitates detection of infection is variable but often transient. Thus, paralytic poliomyelitis is thought to occur in fewer than 1% of children that are infected by the causal enterovirus, and German measles rash (caused by rubella virus, Togaviridae) manifests in about 50% of children and young adults. Comparatively few viruses, such as measles (Paramyxoviridae) and rabies (Rhabdoviridae) seem to be associated with disease in virtually all instances. Isolates attributable to one virus often vary in this as in other respects. Thus, the index isolate of Ebola haemorrhagic fever filovirus from Zaire had one of the greatest known mortality rates, whereas isolates of Ebola filovirus from Sudan has lesser human pathogenicity and the primate relative (Reston) had none. Numerous viruses are not obviously harmful provided that they infect at an early stage in life when the uniquely sophisticated vertebrate defence systems are unprepared. Similarly, plants and insects are more readily infected by viruses when young, demonstrating a phenomenon of mature resistance that is unexplained but probably reflects structural as well as compositional changes in cells/tissues.

Dramatic release and sudden demise of virus-infected cells seems less common in higher plants than in bacteria or animal cells. In many instances, viruses are released by budding (Figure 2.2(a,b)) through boundary membranes of cells in which they have been produced and the cells survive. Probably most virus transport between cells within a host is in the form of nucleoprotein, with or without a lipoprotein envelope, but virus genomic nucleic acid associated with membranes

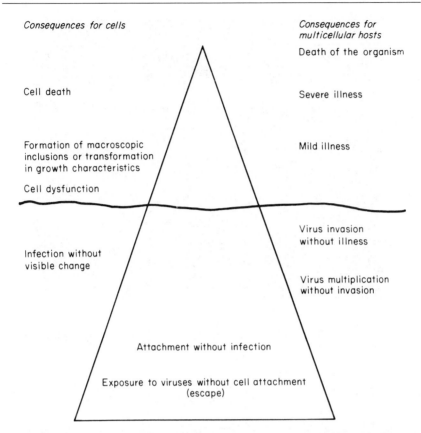

Figure 2.1 The iceberg concept of virus disease at the level of the cell or the multicellular host with varying amounts of discernible effect. After Evans (1982).

may also spread in body fluids to establish remote secondary infection sites.

The speed of cell to cell movement is very slow and few viruses seem to rely on this method of dispersal within their hosts. Herpesviruses and rabies are unusual in that they characteristically invade along nerve axons. The herpesvirus associated with Aujeszky's disease in cattle or pigs localizes in dorsal root ganglia, where its presence is associated with uncontrolled electrical discharge in the nerve endings giving rise to the 'mad itch'. Similarly, the virus of rabies progresses along the neurons to the central nervous system before entering salivary glands from where dissemination may occur when infected animals bite others. The rate of cell–cell invasion computed from measurements of increase of local lesion diameter in leaves of plants is temperature-dependent but about 25 μm/h (Henderson and Cooper, 1977). However, viruses can move at speeds several orders of magnitude greater than

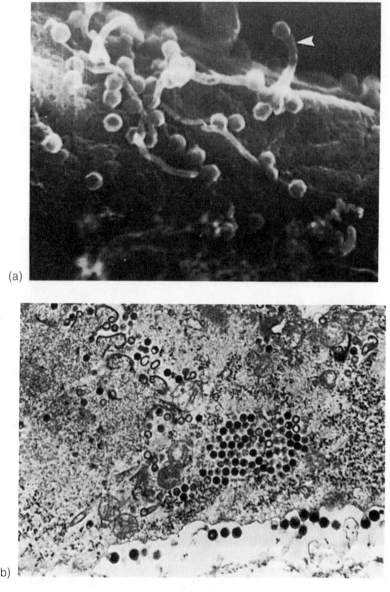

(a)

(b)

Figure 2.2 Transmission (a) and scanning (b) electron micrographs showing an iridovirus, frog virus 3, accumulating within and budding (arrowed) at the plasma membrane of infected chick embryo cells. (×2) Courtesy of D.C. Kelly, NERC, Institute of Virology and Environmental Microbiology.

this within the food-, or less commonly the water-, conducting channels (the phloem or xylem) of plants, and in a similar way, viruses circulate within the blood of vertebrates or in the haemolymph of invertebrates.

2.4 RESPONSE TO THE CHALLENGE

Plant hosts may respond in a variety of ways (Bos, 1970; Matthews, 1980), some of which minimize the rate and extent of invasion by viruses. Hypersensitivity is a type of pathological response which usually takes the form of localized cell death (necrosis) and is often, but not necessarily, associated with restricted virus invasion. Indeed, virus infection may be localized within a plant that does not show symptoms. Vertebrates have been studied more intensively and are known to respond to the challenge of virus inoculation in a multitude of ways (Mims, 1982) including some that invertebrates also have (Chapter 3). Broadly, there are two main types of reaction – cellular and humoral – and both are facilitated by the circulation of blood. In vertebrates, the cellular response is attributed to populations of white blood cells, whereas the humoral resides within the aqueous matrix, the serum.

Bone marrow, thymus-derived lymphocytes, and other white cells (mononuclear phagocytes) engulf virions and thereby modify virus concentration within mammalian blood (invertebrates possess an analogous cellular system). However, virions engulfed by these cells are not necessarily inactivated; even before the wealth of data on human immunodeficiency-associated lentiviruses became available, it was known that viruses remain infectious while within white cells of human blood. In a few instances, togaviruses, such as are associated with Dengue fever, modify directly or indirectly the surface properties of these cells, a phenomenon implicated as a predisposing factor to fatal haemorrhagic shock which follows sequential invasion by related viruses (Halstead, 1970).

Humoral resistance has three functional components: interferons, antibodies and complements. Interferons are a group of glycoproteins produced by nucleated vertebrate cells either as a primary response to virions or double-stranded RNA directly, or, as a second response, by white blood cells following renewed inoculation. These chemicals act as an early line of defence. Although less is known about the phenomenon in plants, it has been proposed but not proved that they also synthesize interferon-like chemicals in response to virus infection (Sela, 1981).

Antibodies are an extremely diverse family of molecules which plants are not known to synthesize. Adaptive responses based on a long-lived immunological memory of lymphocytes is most highly developed in primates, but there are a variety of non-specific responses to challenge by viruses which also limit replication. Three classes of antibody, immunoglobulins G, M and A (IgG, IgM and IgA), neutralize the infectivity of viruses and/or complex with viruses to form clumps (precipitates). Studies with viruses as immunogens (antigens) have shown that the class of antibody and its rate of synthesis depends on the quantity of antigen applied. With low doses of antigen, IgM is produced briefly; with high doses, both IgM and IgG are synthesized (the IgM preceding the IgG), but IgG is usually the most abundant humoral species in blood. A series of IgA-type antibodies also occur in blood but are most concentrated in secretions such as saliva, milk, tears, mucus, etc., where they are well placed to serve a localized protective function. The rate of antibody synthesis is known to be influenced both by environmental and genetic factors. Thus, caucasoid and negroid human populations differ in their rate of IgG metabolism, daily turnover rate being seven times greater among negroids. However, outbreeding populations are non-uniform in their immune responsiveness and the process is also modified by concurrent infection or starvation. Thus infestation with protozoan and helminth parasites, which is almost inevitable, of domesticated and wild livestock diminishes humoral responses on the one hand but enhances cellular resistance systems on the other. Plausibly, malarial infestation of humans (and probably also of non-human primates) contributes to the prevalence of virus infection in tropical regions of the world. A similar synergism has been noted in rodents in which arenaviruses probably cause transient suppression of blood-associated resistance systems in these animals, thereby favouring invasion of the kidneys and salivary glands while facilitating dispersal of the viruses in urine. IgAs secreted at cell surfaces have an impact on viruses infecting at genitourinary or respiratory tracts but viruses introduced directly into blood circumvent this line of defence.

Whereas Ig molecules neutralize free virions, cell-mediated systems eliminate virus from solid tissues (Doherty et al., 1992) but neither is wholly effective in eliminating viruses. Indeed, viruses can replicate in macrophages that have fc receptors on their surfaces and antibody-virus complexes which form as a consequence may facilitate infection by different viruses. Virus infection of lymphocytes or monocytes by Epstein-Barr herpesvirus tends to immortalize these cells and, as with human immunodeficiency lentiviruses which infect both T and B cell lineages, leads to opportunistic infections. Sanctuary from immunological surveillance seems possible when viruses locate in the brain but such tissue tropism is not exercised by many viruses and evasion of antibodies is more usually achieved by antigenic variation which is facilitated when viruses have multipartite genomes. Copying errors during replication constantly give rise to the genetic variation which is sometimes manifested as antigenic drift and characterizes viruses with RNA genomes (Kirkegaard and Baltimore, 1986). Rates of evolution for RNA viruses have been suggested to be one million times greater than for cellular genes. However, selection and survival are sometimes ignored in such guestimates. DNA 'proofreading and repair' does not eliminate errors; indeed, the process seems to be facultative (Parrish *et al.*, 1991).

Other consequences of exposure may include autoimmune diseases. Increasing evidence is linking human immunodeficiency-associated lentiviruses with autoimmune pathology. Somewhat less convincingly, other viruses have been invoked as causes of insulin-dependent diabetes mellitus. Undoubtedly viruses such as Epstein-Barr herpesvirus cause polyclonal activation of B cells and mice cells containing lymphocytic choriomeningitis arenavirus similarly potentiate 'anti-self' responses. Viruses initiate the excretion of novel proteins onto cell surfaces and this has been implicated as a cause of arthritis. Furthermore, viruses 'capture' host genes (Meyers *et al.*, 1989). This process sometimes can be thought to confer selective advantage by minimizing vulnerability to antibody surveillance. However, by expressing immunological features with properties in common with host proteins, the viruses potentially induce 'autobodies' which might be considered to be counterproductive.

A diverse array of viruses, including those pathogenic for plants, insects or bacteria, act as antigens when introduced into vertebrate animals. This phenomenon has been exploited to provide rapid *in vitro* means of virus detection/identification (Weir, 1978; van Regenmortel, 1982; Roitt and Delves, 1992). Additionally, because humoral resistance factors are more or less durable, it is possible to obtain a retrospective picture of vertebrate exposure to (but not necessarily infection by) viruses. In nature, antibodies, complements and interferons have a pivotal role in ecology and evolution of viruses in vertebrates; the defences of domesticated livestock and humans are often deliberately induced by inoculation with viable or inactivated viruses in the form of vaccines (Chapter 6). However, it is sometimes forgotten that vaccines may diminish disease severity but do not eliminate virus replication. Consequently, vaccinated animals augment the diverse range of inapparent infection sources.

2.5 VIRUS TRANSMISSION

2.5.1 VERTICAL TRANSMISSION

When viruses are transmitted from parents to offspring as in pollen, sperm or eggs, the process is described as vertical transmission. Vertical transmission concerns the route of a unit of heredity (which may also be infectious) through families and the term originated because pedigrees are usually depicted with progenitors at the top of a page while descendants occupy positions towards the bottom. The vertical axis in these diagrams, though usually unstated, is time measured in generations. Horizontal transmission involves contemporaries that may be (but need not be) related to one another. In relation to a family tree the direction of horizontal transmission tends to be across a page rather than from top to bottom (Figure 2.3). Many viruses are transmitted both horizontally and vertically (Fine, 1975). Some routes of virus entry into or exit from plants or vertebrate animals are illustrated in Figures 2.4 and 2.5, respectively.

Vertical transmission in vertebrates

The urinogenital tract as a source and portal of entry for viruses in humans received little attention until after antibiotic treatment had diminished the incidence of microbial venereal diseases such as gonorrhoea and syphilis. Contemporaneously, promiscuous human oral and homosexual activities became prevalent and increased the frequency of herpesviruses spread from mouth to genitalia. Herpesvirus contamination of the genital tract challenges the fetus or the human baby during delivery and often causes primary infection with almost invariably fatal result. Furthermore, herpesviruses are associated closely

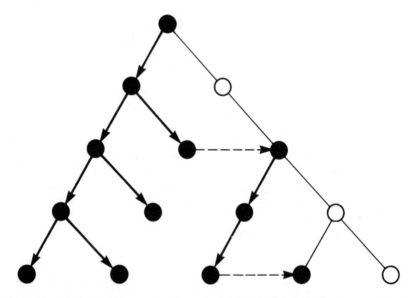

Figure 2.3 Vertical (→) and horizontal (−→) transmission in five successive generations of a population reproducing asexually. Infected individuals are (●) and healthy (○).

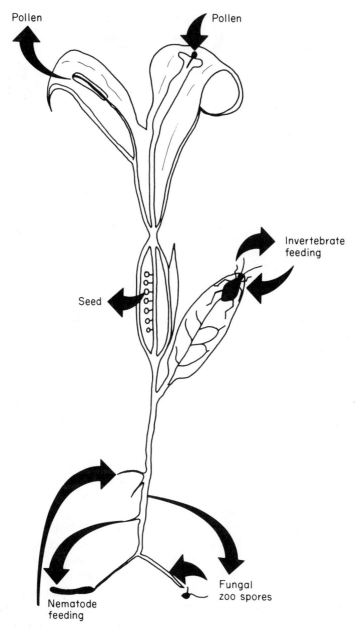

Figure 2.4 Some routes of virus entry into or exit from plants.

with abortion in horses, suggesting that infection may occur similarly in wild and domesticated mammals. However, death is not an inevitable consequence of exposure; the herpesvirus associated with malignant catarrh in wildebeest is probably maintained within herds because

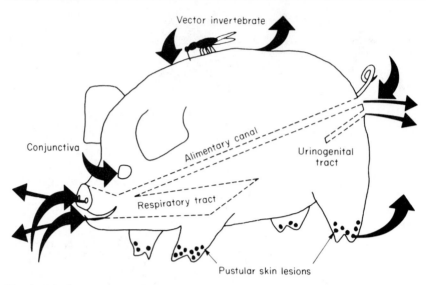

Figure 2.5 Some routes of virus entry into or exit from vertebrates.

intrauterine infection is not fatal. Indeed, the wildebeest calves appear normal yet retain virus in their blood for long periods of time. Domesticated cattle differ in that malignant catarrh is almost invariably fatal for them. The classical association (Gregg, 1941) between malformation of the fetus and maternal rubella (German measles, Togaviridae) during pregnancy is noteworthy but seems somewhat exceptional and is probably irrelevant to the maintenance prospects of the virus. By contrast, the analogous shedding of Newcastle disease virus (NDV) into the yolk of eggs laid by infected poultry undoubtedly contributes (but probably only in a minor way) to the dispersal of vaccine-derived virus isolates that do not severely debilitate their hosts. In eggs laid by hens infected by severely pathogenic isolates of NDV, embryos commonly die within a few days of incubation.

Numerous viruses have been detected in ejaculates but the association seems incidental in most instances. However, somatic cells sometimes fuse with spermatozoa and any viruses present might infect in this way. Excepting the retroviruses associated with immunodeficiency or leukosis/sarcoma diseases of poultry for which most significant transmission is from parent to offspring, sperm is probably an unimportant means of virus inoculation in nature. Nevertheless, care must be taken to eliminate the virus contamination of semen used for artificial insemination; pestiviruses of ungulates, papilloma, parvoviruses and some herpes serotypes, have long been known or suspected to invade from such inocula.

The occurrence of viruses in urine (viruria) is commonplace in mammals; for adenoviruses, such as cause canine hepatitis, and are-naviruses, causing Bolivian haemorrhagic fever or lymphccytic choriomeningitis (Ackerman *et al.*, 1964), urine seems important in maintaining the chain of infection (vertically and horizontally) in rodents. Additionally, urine contaminates food and water thereby facilitating human infection (horizontally).

Vertical transmission in seed plants

For many years, the transmission of viruses via seeds of their plant hosts was considered a rare curiosity and the importance in virus ecology was consequently trivialized. By contrast, it is now suspected that some viruses exist, and the distribution of others is greatly facil-itated because they are seed-borne (Mink, 1993). Although the literature on the subject is confused by a mass of data concerning inadequately described agents of disease (Mandahar, in Maramorosch and Harris, 1979) and all virus–host combinations have not been tested equally, more than 100 viruses (representing many ICTV-approved taxa) have been reportedly transmitted in seeds. Knowledge of the hosts in which viruses are seed-borne is predictably uneven; most is known about a few of the agriculturally important species of the Chenopodiaceae, Graminae, Leguminosae, Rosaceae and the Solanaceae. Even though the complete absence of seed transmission is unrealistically difficult to prove, it is noteworthy that a diverse collection of viruses that have been tested seem not to be transmitted in seed of their hosts. Thus, none of the agents that replicate in vascular tissue (e.g. geminiviruses) or those which are largely restricted to vascular tissue (e.g. luteoviruses in the absence of umbraviruses) or those that typically infect only locally (e.g. tobacco necrosis necrovirus) are seed-transmitted. Further-more, with one notable exception (peanut clump furovirus in *Arachis*), no virus that is known to have a fungus vector is also seed-borne. However, vertical transmission is normal for cryptoviruses (Boccardo *et al.*, 1987) and prevalent among ilarviruses, hordeiviruses, the non-circulative aphid-borne potyviruses and cucumoviruses, the tospo-viruses, nepoviruses and one subset of the tobraviruses.

For many viruses, the frequency of seed transmission varies markedly with the host genotype. Thus, the incidence of soybean mosaic potyvirus reportedly varies from zero to 70% depending on the soybean genotype involved. Furthermore, virus isolates also differ; some seem to have a predilection for seed transmission. Thus, in barley, many barley stripe mosaic hordeivirus isolates are not seed-transmitted, whereas others are carried by more than 50% of the progeny. Similarly, the tobraviruses can be divided into two subgroups: pea early browning viruses (PEBV) which naturally infect cultivated peas, but not weeds associated with infected crops, and tobacco rattle viruses (TRV) which have very many natural hosts including numerous weeds. Whereas both subgroups are transmitted by trichodorid nematodes (Chapter 3), it is tempting to infer that because seed transmission in peas determines survival, this has tended to stabilize (by inbreeding) virus isolates presently attributable to PEBV.

Transmission of a virus to all the progeny of a host seems uncommon but has been recorded for cryptoviruses and the tobamovirus which naturally infects a geographically isolated population of *Nicotiana velutina* in Australia. In many instances, transmission through 60–90% seed has been observed with the frequency of transmission tending to vary inversely with the time that the viruses are retained by their vectors. Very small frequencies of transmission can have profound epidemiological consequences. Vanishingly small frequencies (2/4000) of seed transmission (as with zuccini yellow mosaic or lettuce mosaic potyviruses) provide numerous foci of infection and potent sources of inocula when aphid vectors of these non-persistent viruses are abundant (Schrijnwerkers *et al.*, 1991).

Virus infection in seed may have several consequences. Sometimes there is a direct effect on components of yield: the cucumovirus of tomato aspermy (as the name implies) causes sterility. However, less drastic effects are normal. Thus soybean mosaic potyvirus diminishes seed weight by 10–20% while additionally being associated with a discoloration of the seed coat, which lessens marketable value. Although there are many records of seedlings derived from infected seed being visually unimpaired, there are almost as many showing that seedlings grown from infected seed are less vigorous. This is probably one manifestation of non-uniform host genotype that complicates prediction about the prospect of virus survival and dispersal in wildlife. However, when aiming for improved crop vigour, plant breeders may unknowingly select for virus tolerance, thereby increasing the likelihood of virus dispersal. Infected plants of the newly selected genotypes are less likely to be shaded out by adjoining healthy plants and incidentally provide exposed foliage on which transient aerial vectors may feed. Additionally, the vigorous virus-infected plants may be able, independently, to perpetuate the virus because they are likely to flower and set fruit. Presumably because of effects on rate of virus replication as well as host metabolism, the amounts of seed transmission vary in response to environmental factors, such as temperature before flowering, although different viruses in the same host may be differently affected by temperature. Indeed, the geographic distribution of some nepoviruses (e.g. raspberry ringspot and strawberry latent ringspot) but not others (e.g. arabis mosaic) seems to be attributable at least in part to adaptation for seed transmission in natural hosts (e.g. chickweed,

Stellaria media), under specific climatic/temperature conditions (Hanada and Harrison, 1977).

In his comprehensive and perceptive analysis of the literature, Bennett (1969) identified two main sorts of seed transmission: contamination and a more deep-seated infection. The former seems characteristic of stable viruses that reach high concentrations in hosts that they invade systematically. Thus tobacco mosaic tobamovirus is seed-borne in tomato and pepper but is more or less confined to seed coats liable to abrade the first seedling leaves (particularly during transplantation), thereby effecting inoculation.

Although some bean seeds exposed to virus-contaminated soil acquired southern bean mosaic sobemovirus (presumably through cracks in the seed coat) and produced infected seedlings, embryo infection is, in most instances, an essential pre-requisite for 'seed' transmission. Factors influencing the speed of establishment in metabolically active tissues presumably determine its occurrence and frequency of seed transmission. Details about the process are only now becoming known; in one instance a virus (pea seed-borne mosaic potyvirus) seemed to regulate seed metabolism, thereby favouring invasion (Wang and Maule, 1995). Recognizing that six viroids are seed-borne (although two others are not), it is possible that the seed-infecting agent may in some instances be in the form of RNA rather than nucleoprotein particles (Mink, 1993). Some viruses infect embryos but do not persist, infectivity being lost as the seed matures. However, there is little evidence suggesting that differences in frequency of seed transmission reflect differences in this presumed inactivation; the rate of virus (infectivity) loss is usually less than the rate at which seeds lose viability. Nevertheless, there are records of viruses remaining infective in seed that germinated after 14 years' storage.

Embryos can be infected via two routes: from the enveloping maternal tissue (the megagametophyte) and from the microgametophyte during fertilization. The relative contribution of the parents varies (Figure 2.6). Thus, with cherry leafroll nepovirus (CLRV) in birch, embryo viability depends on whether the virus is present in the male, the female or both parents (Cooper *et al.*, 1984).

In raspberries, Murant and Lister (1967) observed that pollen carrying raspberry ringspot nepovirus competed poorly with virus-free pollen when both were present together at stigmatic surfaces and concluded that in this situation pollen transmission to ovules would have little importance. Their experiments were the best controlled (against genetic traits) but the data may not be applicable outwith this commercial crop. There are few critical data but it seems that viruses which have efficient alternative means of dispersal are more likely to be associated with pollen abnormalities than those that have no known auxiliary method

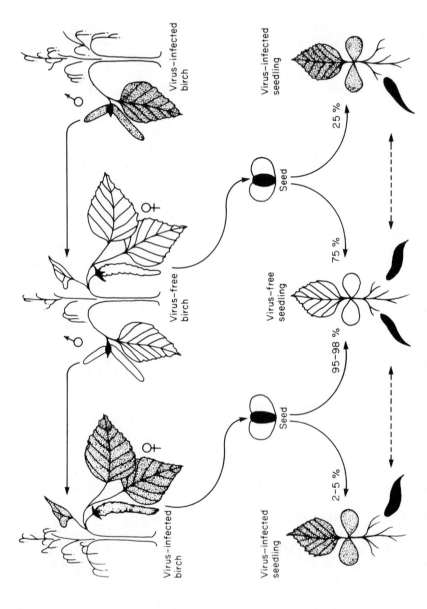

Figure 2.6 Differing contribution to seed of virus (cherry leaf roll) from birch pollen or infected mother trees. Nematodes possibly also transmit this virus between birches and other genera.

of spreading between plants such as barley stripe mosaic hordeivirus (BSMV). Indeed, even though BSMV has been associated with abnormal anther dehiscence, the quality of virus-carrying pollen seemed unimpaired (Carroll, in Maramorosch and Harris, 1979). The literature and general implications of pollen–virus association was critically reviewed by Cooper *et al.* (1988). In the specific context of birch, cherry and walnut, Massalski and Cooper (1984) experimentally determined features of CLRV-pollen associations and used these experiences when modelling its relevance to vertical virus transmission (Cooper *et al.*, 1984).

Pollen can inoculate the gametophyte (giving infected seed) but can also inoculate the mother plant, a phenomenon characteristic of ilarviruses (Mink, 1992). As indicated in Chapter 3, ilarvirus transmission in association with pollen is currently attributed to the activities of thrips. Viruses that reach high concentration in their hosts typically contaminate pollen surfaces and, in hosts such as birch and walnut, in which pollen tubes penetrate through integuments around the ovule (rather than via the micropore between them), the process may facilitate inoculation of maternal tissue. Virus replication within pollen grains, which might be a necessary prelude to maternal infection, has not yet been studied in detail but electron microscopy has revealed accumulations of particles resembling BSMV or CLRV (Figure 2.7) within the cell cytoplasm and nucleus of pollen grains.

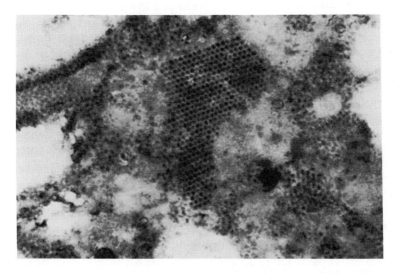

Figure 2.7 An electron micrograph of a paracrystalline array of spherical virus-like particles in pollen grain of birch (×62 500). Cherry leaf roll virus was detected in similar pollen. Courtesy of NERC, Institute of Virology.

Vertical transmission in lower plants

Because of commercial sensitivities about intellectual property and processes, details concerning the impact of phages on industrial reactions are not readily available. However, the potential impact of such viruses is likely to be in the vanguard of concerns about the use of genetically manipulated bacteria for the production of, for example, bovine somatotropin, insulin, blood clotting factors, etc. In more traditional fermentation-based food technologies (for milk, bread, wine, etc.) reliance on a single starter culture genotype makes the process vulnerable to viral impact and calls for expensive design and operations in addition to environmental monitoring.

Transmission of mushroom viruses through infected spores is epidemiologically important and spores (whether they are the products of quasi-sexual or sexual reproduction) are probably crucial in the dissemination and maintenance of viruses in diverse fungal populations. Similarly, algae, protozoa and bacteria are known, or strongly suspected, to contain viruses that persist in populations largely because their genomes are capable of integrating within those of their hosts. With the recognition that algae can provide new sources of valuable chemicals for the pharmaceutical trade or provide food supplements, there has been some research interest and virus-like particles have been noted in about 50 eukaryotic algal types spanning most of the classes within the group (van Etten et al., 1991). Because of difficulties in algal culture, few host–virus systems have been studied in detail but, having regard to the parallel patterns of distribution and exploitation of aqueous environments, it is becoming apparent that algae share with bacteria and fungi a marked tendency towards vertical virus transmission. Thus, Chlorella spp. which can form symbiotic associations with Paramoecium are known to support viruses which are expressed at only one stage in their life and cells which contain these virions do not lyse. Analogously, virus infections, prevalent (c. 3%) in a marine brown alga (Ectocarpus spp.) in coastal regions from Ireland to California, Chile, New Zealand and Australia, express themselves intermittently; in this instance only in reproductive thalli (Muller et al., 1990). It is known that symbiosis with Chlorella favours the survival of Paramoecium and that the microalga only becomes infected by viruses after release from the Paramoecium. However, it is not known whether this is a phase change in the virus population which is normally lysogenic or whether it reflects a carrier state (in which only a small part of any population is constantly infected) which occasionally becomes lytic. Even though eukaryotic algae are crucial components in freshwater and marine environments, there has been little study of the impact of viruses on population structure or system integrity. However, viruses have been invoked in population cycles and spring blooms of diatoms – possibly indirectly by virtue of phage-induced lysis of bacteria, thereby providing food for some components of the food web while diminishing predators from the food chain (Bratbak *et al.*, 1990). So far, viruses lytic for indicator algal cultures have not been recorded in freshwater, although viruses capable of infecting, for example *Chlorella*, are almost universally present (Zhang *et al.*, 1988).

The phenomena associated with vertical virus transmission in bacteria has given rise to a specialized terminology. 'Lysogeny' is a recombination/excision phenomenon which has been extensively studied in the context of the bacteriophage lambda (a phage which is described as 'temperate' because some infected bacteria survive). This contrasts with the behaviour of 'virulent' or 'lytic' phages which typically destroy their hosts when releasing the viruses. Survivors carry the lambda viral genome as a stable 'prophage' in the host chromosome where it is maintained from generation to generation (Ptashne, 1992). The lysogenic state can be interpreted as a means of survival for viral genomes which might be beneficial in circumstances when the phage is more sensitive to environmental stress than its host (not an inevitable state). However, the process more importantly co-locates viruses with potential hosts. Thus starvation/slow growth of bacterial populations tends to lead to lysogenization whereas infection at times of rapid bacterial multiplication tends to result in lysis, thereby diminishing populations of potential hosts which feeds back and constrains the lytic cycle. Supporting the parasitic virus is costly but the specific enzymology which the virus uses for recombination/excision also has a cost (and it is the phage which largely determines the outcome): the virus tends to actively repress phage expression until it is appropriate to make a change. The integration and excision of the lambda phage genome is substantially more efficient than the intrinsic systems of the bacterial hosts which also effect recombination.

Archaebacteria are distinct from bacteria and from eukaryotes and have a notable ability to tolerate extreme environments, such as saltiness, sulphur, heat and anaerobiosis. In step with their increasing importance as sources of industrially important enzymes, more is being discovered about the viruses which infect archaebacteria but ignorance of efficient inoculation and bioassay processes delays investigation. Until recently, the most notable feature of virus-like particles in archaebacteria was their bizarre lemon shape (in sulphobacteria) and their propensity to cause lysis. However, a few other interesting observations have been made on interactions involving archaebacteria, their viruses and their environment (Reiter *et al.*, 1988; Schleper *et al.*, 1992). Thus, the phage titre in a natural brine pool was noted to vary inversely with salinity; at saturated salt conditions, the archaebacterium species concerned attained a carrier state with very low virion liberation and only a tiny part of the population was infected. By contrast, when the salt in the pool was diluted by rain (a potential threat to host survival) the lysogenic state was terminated, the bacteria burst and liberated 'mature' virions which can be considered as stable structures better able to 'survive' adversity to which their hosts might be vulnerable. Another archaebacterium, the sulphur-dependent thermophile, *Thermoproteus tenax*, reacts analogously to sulphur concentration. Thus, when sulphur is depleted and death of the archaebacterial host is threatening, the virus changes from its normal lysogenic state of integration, multiplies and escapes by lysis.

Molecular probes show that natural populations of bacteria and archaebacteria retain prophages; indeed, in *E. coli* K-12, about 3% of the bacterial genome is attributable to integrated phage lambda. Prophages sometimes determine the phenotype of their hosts as when they modify polysaccharide surface antigens, rendering the bacteria resistant to extraneous phage attack. Furthermore, phage genes directly impact upon the pathogenicity of *Neisseria/Clostridia* responsible for diphtheria, botulism or toxic shock in vertebrates (Betley *et al.*, 1992). Toxaemia is deleterious to the survival of the host and therefore the bacteria and the phages. Explanations for the paradox might be, as with phage lambda, an incidental resistance to phagocytosis: alternatively the genes for toxin production are recently acquired.

Totiviruses, which are among the viruses that are known to be vertically transmitted in filamentous fungi or protozoa, illustrate the adaptive benefits which can be associated with infection – in yeasts totiviruses facilitate excretion of protein toxins which are harmful to competitors. This 'killer' trait has been used in the brewing industry to minimize the inconvenience due to extraneous populations of yeast. Analogously, fungal cultures infected with partiviruses and totiviruses are hypovirulent (Brasier, 1990) and, to minimize damage from chestnut blight in the USA, cereal 'take-all' or Dutch elm disease in UK, the release of virus-infected fungal cultures has been contemplated.

2.5.2 CONTACT SPREAD IN PLANTS

Whereas the physicochemical processes that facilitate membrane penetration by viruses are similar in plant and animal cells, plant cells differ fundamentally in being bound by rigid cellulose walls. Furthermore, the greater part of a plant's external surface is covered by hydrophobic secretions (cutin and/or waxes) impervious to viruses. Although some types of surface cell (e.g. stomatal guard cells, root/leaf hairs which allow controlled nutriment exchange in plants) lack thick coverings and therefore seem especially vulnerable, wounds nullifying the physical barriers are, with few exceptions, essential prerequisites to plant infection. Normally, wounding is achieved by the feeding of another organism (e.g. insect, nematode, fungus) which has a specific, albeit sometimes transient, relationship with a virus that it can introduce to initiate an infection. Additionally, there are a few records suggesting that virus inoculation of plants occurs more or less casually when leaf miner flies lay eggs in foliage, when aphids walk over leaves or when the following feed: slugs, snails, grasshoppers, larvae of butterflies, moths and leather jackets.

Because the plant cell wall is permeated by cytoplasm (ectodesmata/plasmodesmata) and because the frequency of virus infection in solanaceous hosts varies with the prevalence of ectodesmata in the outer wall of epidermal cells, it is possible that erosion of the superficial secretions by abrasion may alone be enough to allow virus access to cytoplasm. In nature, leaves of virus-infected plants repeatedly abrade neighbouring foliage, breaking hair cells and thereby providing portals of entry on the one hand and inoculum on the other. Similar things happen below ground. As roots grow through soil, abrasion results in damage. In cultivated crops, the prospects for virus dissemination are greater than in natural communities because infectibles are likely to adjoin hosts, and damage is an inescapable consequence of crop management, i.e. weeding, harvesting, cultivating, inspection, etc. Viruses that spread to a significant extent by contact are usually stable and typically achieve high concentrations in plants that they infect. Thus, tobacco mosaic tobamovirus which is spread by contact in tomato and tobacco crops retains infectivity in non-sterile extracts for 50 years and readily contaminates hands, clothing and implements. Tobamoviruses and others (e.g. potexviruses) that can spread by contact sometimes spread in the absence of leaf contact and this has been interpreted in terms of mechanical inoculation underground (by root contact), the presence of a soil-inhabiting vector organism or transmission in airborne debris (Allen, 1981). Inevitably, damage to the aerial parts of plants has implications for roots because viruses tend to be washed into the soil. Some of this virus will be metabolized by microorganisms or adsorbed to silt or clay, and its role as inoculum is probably negligible thereafter (Cheo, 1980). However, there is an accumulating body of evidence showing that infectious viruses (notably tombusviruses and necroviruses) occur in surface waters (Koenig, 1986). Initially it was assumed that virus was released only as a consequence of root injury or decay but it is now known that undisturbed and unblemished roots of systematically invaded hosts growing in nutrient solutions release viruses in the absence of abrasion, microbial degradation or virus-induced cell death (necrosis). This release has been recognized in a variety of virus–host systems but it is difficult to assess the biological importance of the phenomenon. Probably the amount of virus liberated from intact roots is trivial when compared with that resulting from disintegration of infected host tissues. Nevertheless, soil contaminated with debris from infected stock is a potential source of inoculum for glasshouse-grown crops (notably in hydroponic systems) which probably become infected after mechanical injury to root hairs or because of fungal vectors.

The fact that a diverse array of viruses may be released into soil implies that casual root infection occurs occasionally, even when the viruses concerned are normally transmitted when pathogens infect or pests feed. The mechanism of release is uncertain but the sources are important for some fungal vectors of plant pathogenic viruses.

2.5.3 VIRUS TRANSMISSION BY FUNGI

Although fungi have not yet been implicated as vectors of viruses pathogenic for insects or for other animals, it would not be surprising if this occurred. During the past 20 years, fungi have been recognized as hosts (Hollings, 1978; Lemke, 1979) and as vectors of plant pathogenic viruses which may be divided between four groups (Teakle, in Kado and Agrawal, 1972; Cooper and Asher, 1988, Adams, 1991). Notwithstanding the claim that tobamoviruses may be carried between plants by the oak powdery mildew fungus and the claim that rust fungi are virus vectors, two classes of obligate root parasite are well-documented: *Chytridiales* (e.g. *Olpidium* spp.) and *Plasmodiophorales* (e.g. *Polymyxa* and *Spongospora* spp.). Both types of fungus have motile zoospores that migrate between roots of flowering plants in films of soil water, attach themselves at new locations, withdraw their flagellae and penetrate the superficial host cells in which a new thallus forms for a few days before the cycle is repeated.

During their brief period of freedom (1–2 hours), the total zoospore surface is liable to become contaminated with virus particles from the soil water. Tobacco necrosis necrovirus, which infects species in some 40 dicotyledonous and monocotyledonous families, but causes economic disease in only a few cultivars of tulip and potato, is transmitted in this way by *Olpidium* spp. There is considerable specificity reflecting surface charge properties between isolates of the viruses and their vector zoospores (Mowat, 1968). Furoviruses (e.g. soil-borne wheat mosaic, potato mop-top, beet necrotic yellow-vein) have permanent relationships with their plasmodiophomycete vectors (species of *Polymyxa* and *Spongospora*) and a similar persistent relationship exists between a distinct collection of bymoviruses (causing barley yellow mosaic, barley mild mosaic, rice necrosis mosaic, wheat yellow mosaic and wheat spindle streak

(Reichmann *et al.*, 1992; Shukla *et al.*, 1994). Plasmodiophoromycetes acquire their ability to transmit both sorts of viruses while growing within cells, not by virus adsorption to zoospores in the soil water. Zoospores of *Polymyxa betae* carrying beet necrotic yellow-vein furovirus contain virus-like particles, but although it is plausible, there is as yet no direct evidence that any of these viruses multiply in both the fungal vectors and in flowering plants.

2.5.4 CONTACT SPREAD OF VIRUSES BETWEEN VERTEBRATES

The vertebrate body presents three large moist surfaces to the environment, the skin, the respiratory mucosae and the gastrointestinal lining, plus two lesser surfaces, the eye and the urinogenital tract. Each is an important infection court for vertebrates and, although less is known about invertebrates, the alimentary canal seems to be the most important site of inoculation in these animals also.

Now that smallpox has been virtually eliminated, the skin lesions attributable to herpesviruses or to papovaviruses (which cause warts) are the main examples of human contact-spread viruses. There is an association between ulcerative conditions of all sorts with infections by hepatitis B or human immunodeficiency lentiviruses but blisters or pustular lesions associated with myxomatosis in rabbits, aphteroviruses in cattle or caliciviruses in swine provide sources of contagion, as do dried fowlpox (Poxviridae) scabs or contaminated feather follicles in Marek's Disease (Herpesviridae). Viruses discharged from avian hosts in these ways remain infectious for 2–3 weeks, during which time dried faeces and contaminated feathers may also become air-borne.

Being a normal function of life, respiration is one of the most common methods for the intake and output of virus-laden air. In addition to respiration, many animals (e.g. dogs, swine) use their noses to investigate their environment, thereby increasing their exposure to inhaled infectious material. The numerous human rhinoviruses (Picornaviridae) associated with the common cold disease, influenza A, B and C (Orthomyxoviridae), respiratory syncytial virus (Paramyxoviridae) and the normal infections of childhood (measles and mumps) are all spread in air. Wild and domesticated animals also have their influenza, bronchitis and pneumonia viruses as well as canine distemper, rinderpest of cattle, porcine, canine and feline parvoviruses, etc. It was at first thought that foot-and-mouth disease aphthovirus (FMDV) entered the blood of animals after ingestion and via lesions in the alimentary tracts. However, it is now generally accepted that most natural infection takes place after inhalation of aerosols containing virus. Indeed, the pharynx seems to be the prime source of FMDV in wild buffalo, although additional sources of viruses in aerosol form are undoubtedly generated from lesions on the mouth, feet, teats, and via milk and faeces.

2.5.5 VIRUSES IN FOOD

All natural foods may contain viruses and the exploitation of hitherto unused sources of food prepared in novel ways, from time to time, reveals unsuspected pathogens. Vesicular exanthema in pigs was almost certainly a result of feeding flesh from beached marine mammals and thereafter feeding undercooked pork to other pigs. Furthermore, changed processing systems applied to offal and 'recovered meat' intimately associated with bones has been implicated in the transfer across species boundaries of agents (prions) of infectious dementia .

Parasitism/predation routinely facilitates virus dispersal on the one hand and inoculation on the other. Invertebrate pests of plants or animals are notable for their ability to transmit viruses that they acquire while feeding. In some instances the invertebrates are merely contaminated with viruses and the association is ephemeral but in others the invertebrates are infected and retain a life-long ability to inoculate (Chapter 3). Vertebrates act as virus vectors less commonly, although they undoubtedly passively transport diverse invertebrates which themselves carry viruses. However, within populations of wild carnivores, such as skunks, foxes, hyenas, mongooses, jackals, wolves and, in the New World, also vampire and insectivorous bats (Figure 2.8), biting is an important way in which rabies-like viruses (Rhabdoviridae) are spread (Bisseru, 1972). Furthermore, some togaviruses seem to circulate within populations of bats because hosts bite infectibles. A different range of viruses that remain infectious after ingestion and passage through the alimentary canal are also spread by predators. Enteric picornaviruses and reoviruses of vertebrates typically replicate in gut cells of their hosts and are liberated in faeces, which contributes to the contamination of water (Chapter 5). Some of these viruses are also spread when the mouth parts of detritivors become contaminated or because of predation. Viruses pathogenic for invertebrates, such as the baculoviruses causing nuclear polyhedrosis of *Gilpinia hercynia* sawflies, are dispersed when birds feed on larval cadavers that contained virus at their death. The birds deposit viruses (with faeces) on to leaves where they are liable to be eaten by other larvae which become infected as a consequence. Indeed, ingestion with food is the best known way in which Hymenoptera, Lepidoptera and Coleoptera become naturally infected with the viruses which affect them. Caterpillars and mosquito larvae are in some instances infected as a consequence of cannibalistic behaviour, and insect larvae that eat

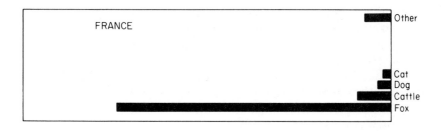

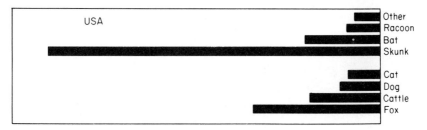

Figure 2.8 Rabies, relative numbers of confirmed animal cases in Europe and North America where skunk rather than fox is most commonly affected. From Office of Health Economics.

their way through egg cases contaminated with baculoviruses become infected in this way.

Within hives, viruses spread because infected bees exude viruses from glands that also secrete compounds which the bees add to the pollen they use as food. Interestingly much of the honey, and the pollen that bees collect, which is used for human consumption contains viruses infectious for bees (Bailey, 1981) or plants (Mink, 1983). As far as is known, vertebrate pathogens are not included.

Secretions from vertebrates also contain viruses but there must be a very high level of viraemia to produce sufficient inoculum to cause infection. Nevertheless, milk is a potentially significant means of vertical and perhaps also horizontal transmission of parvoviruses and retroviruses in populations of wild animals such as rodents. The antigens associated with hepatitis B have been detected on a few occasions in human milk and goat's milk has been implicated as a source of inoculum for togaviruses such as cause encephalitis in Central Europe. Milk from cows, goats, etc. may, in rare instances, contribute immunizing doses of viruses to human drinkers. However, in countries where milk is normally heat-treated (pasteurized) before consumption, it seems to be an unimportant source of viruses.

Although most viruses are eliminated by post-mortem changes in their former hosts, a significant few retain their infectivity even after carcasses are frozen; Newcastle disease paramyxovirus retains its infectivity for 250 days in poultry stored at −20°C. Furthermore, uncooked meat, bones and offal have been implicated as the probable factors responsible for the spread of the calicivirus causing rabbit haemorrhagic disease and for re-introducing the viruses causing swine vesicular disease into the USA or foot-and-mouth diseases into the UK, where a rigorous exclusion and slaughter regimen keep the islands substantially free of these agents.

Viruses associated with invertebrates

3

3.1 INTRODUCTION

Numerous viruses pathogenic for plants or vertebrates are dispersed by invertebrates that they contaminate incidentally and more or less transiently. Furthermore, some evidence suggests that endoparasitic wasps (Hymenoptera) transmit insect pathogenic ascoviruses (Govindarajan and Frederici, 1990) and conceivably are thereby responsible for pest suppression where insecticidal chemicals (which would incidentally kill the vector wasps) are not used (Fleming, 1992). An additional range of rhabdoviruses, reoviruses and togaviruses are transmitted by invertebrates that are themselves infectible by the agents which they acquire from and inoculate into plants or vertebrates. Despite chronic and persistent infection, these vectors usually appear normal. Inapparent infection, though common among invertebrates, is however not inevitable. Viruses have long been implicated as causes of death/diseases affecting insects reared for man's benefit (e.g. silkworm, *Bombyx mori*, or bees, *Apis* spp.) and, in recent years, crop pests have been investigated intensively to assess the commercial potential of pathogenic viruses as natural insecticides. A variety of microbial agents offer the prospect of augmentation or replacement of chemical pesticides (Fuxa and Tanada, 1987). Although many viruses have the ability to replicate in invertebrates including aphids (Gildow and D'arcy, 1988), most have close relatives which are vertebrate pathogens or have some other negative feature associated with them. The most promising and researched viruses which seem arthropod specific are baculoviruses (Payne, 1988; Wood and Granados, 1991). Despite the absence of known hazard for vertebrates, baculoviruses are slow acting, have low efficacy against late instar larvae (that tend to eat most) and are sensitive to solar radiation. Engineering of baculoviruses to improve efficacy has attempted to incorporate hormones that interfere with insect develop-

ment/behaviour, growth and water regulation (Maeda, 1989; Bonning
et al., 1992). Among the alternative approaches which are under
investigation are the expression of toxins following gene incorporation
in the baculovirus genome, e.g. *Bacillus thuringiensis* delta endotoxin
(Merryweather *et al.*, 1990) or scorpion venom (Stewart *et al.*, 1991).
Theoretically, engineering a toxin gene into a baculovirus improves the
specificity of delivery since expression depends on ingestion and replication in a permissive host and, against a background of controversy, a
number of genetically engineered baculoviruses have been used in
authorized releases into the environment (Cory *et al.*, 1994).

3.2 INVERTEBRATES AS VIRUS VECTORS

No vectors are known for more than 500 virus-like agents that affect
plants or vertebrates and it is probable that hitherto unsuspected sorts
of vectors will soon be recognized. Indeed, Foil and Issel (1991) argued
that, because retroviruses mutate rapidly and occur in blood where
they are available to haematophagous arthropods, it was only a matter
of time before the retroviruses were selected for their ability to replicate
in insect cells.

Plants seem particularly prone to infection by viruses carried by
homopterous insects (notably aphids, whiteflies and hoppers), whereas
haematophagous dipterans (notably mosquitoes) seem the most
important vectors of viruses pathogenic for birds and other vertebrates
(Table 3.1). Although these sorts of insects collectively transmit several
hundred viruses, biologically significant vectors also occur in the other
invertebrate groups. Together, ticks and mites (Acarina) transmit
dozens of viruses. About 20 plant pathogenic viruses have nematode
vectors and similar invertebrates may perhaps also facilitate dispersal
of one or two viruses that are vertebrate pathogens.

Invertebrate vectors have a variety of life cycles and habits but are
unified by their ability to pierce cells, and imbibe the contents of plants
or higher animals on which they feed. In subsequent sections a few
details are given about the habits of vectors because of their relevance
to the understanding of virus ecology. Arguably, most is known about
aphid associations with some of the economically important plant
pathogenic viruses they carry. Unfortunately, much of the mass of
knowledge concerning homopteran vectors that has accumulated since
1895 (when Takata first implicated leafhoppers as vectors of the agent
causing rice dwarf disease; a phytoreovirus) is equivocal and interpretation is often difficult. A great deal of information concerned with
insect transmission of plant pathogenic viruses has been collated in the
multi-authored books edited by Gibbs (1973), Mayo and Harrap (1984),

Table 3.1 Some invertebrate taxa that facilitate the dispersal of viruses pathogenic for plants or vertebrates

	Taxonomic group	Relative importance for plant pathogenic viruses (and relationship*)	Relative importance for vertebrate pathogens (and relationship*)
Insecta			
	Diptera	± (M)	++++ (M,P)
	Hemiptera (Sternorrhyncha)		
	Aphididae	+++ (M,C,P)	–
	Pseudococcidae	+ (M,?C)	–
	Aleyrodidae	++ (M,P)	–
	Hemiptera (Auchenorrhyncha)	+++ (M,P)	–
	Heteroptera	+ (P)	–
	Coleoptera	++ (M,C)	–
	Thysanoptera	+ (C,?P)	–
	Orthoptera	± (M)	–
	Lepidoptera	± (M)	–
Acarina			
	Eriophyidae	++ (M,?P)	–
	Tetranychidae	± (M)	–
	Ixodidae	–	+++ (P)
	Argasidae	–	+ (P)
Nematoda		++(M)	± (?)
Gastropoda		± (M)	–

* M, mechanical, i.e. non-circulative contamination.
C, circulative, i.e. virus enters haemolymph but is not known to multiply in vectors.
P, propagative, i.e. multiplies in vectors.

Maramorosch and Harris (1979, 1981) and Harris and Maramorosch (1977).

3.3 NEMATODES AS VIRUS VECTORS

Undoubtedly some helminths cause wounds in the alimentary canals of their vertebrate hosts and, although unproven, nematodes have been implicated as possible vectors of a few viruses pathogenic for animals (e.g. ascarids for the rhabdovirus of vesicular stomatitis and lungworms, *Metastrongylus elongatus*, for the paramyxovirus causing influenza in pigs). Sparse evidence (Shope, 1955) also suggests that *Metastrongylus* eggs contain virus acquired from swine with influenza. If confirmed this might imply perennation, and conceivably these worms may be capable of transmitting the virus back to pigs that ingest the eggs. There is, however, no evidence suggesting that the virus replicates in *M. elongatus* and it is doubtful if helminths are important agents either of dispersal or perennation of viruses pathogenic for vertebrates. By contrast, nematodes are essential vectors for some plant pathogenic viruses (Lamberti *et al.*, 1975).

Plant parasitic viruses with nematode vectors were first recognized in Western Europe and North America but more recent observations suggest that they occur in most developed countries in temperate and tropical regions of the world. Greatest damage from feeding is attributable to plant parasitic nematodes in the order Tylenchida. The nematode vectors are distinct and are included in the Dorylaimida and in the families Longidoridae or the Trichodoridae. Whereas longidorids (*Xiphinema* and *Longidorus* spp.) are 2–12 mm long as adults and have hollow spears that are used to pierce and as feeding tubes to extract contents of root cells, trichodorids (*Trichodorus* and *Paratrichodorus* spp.) are shorter (0.5–1.5 mm long as adults) and feed more superficially, often in epidermal cells which are pierced by a solid curved murine tooth (onchiostyle). Individually, longidorids are the more destructive feeders (their spears commonly penetrating through cells to the vascular tissue). However, when feeding, nematodes in both groups inject glandular secretions into cells and associated histological changes give rise to swellings (galls or stubby roots). Having found a root, nematodes explore the surface before attempting penetration (which is not always achieved) and feed only intermittently, sometimes requiring days of access for virus acquisition and transmission. Data vary for these and other technical reasons, but *X. americanum* can acquire an ability to transmit tomato ringspot nepovirus within 60 minutes of access, and *Paratrichodorus allius* acquires and may transmit tobacco rattle tobravirus in a similar period. Inoculation probably occurs in a single brief probe and there is no latent period before transmission. In groups of vector nematodes that have acquired virus, transmission rate diminishes during starvation; *Xiphinema* spp. and trichodorids tend to retain an ability to transmit 'their' viruses for longer periods (10 months) than do *Longidorus* spp. (days to weeks). There is little information, but the feeding of nematodes in virus vector genera is probably not restricted to flowering plants; bryophytes, fungi and algae may also be fed upon. Nematode-transmitted plant pathogenic viruses are carried as extracellular contaminants, do not persist after moulting and are not known to pass to progeny via eggs. In longidorids, electron microscopy has revealed virus-like particles (VLPs) in the gut lumen between the spear and its sheathing cuticle or within the hollow spear itself (Figure 3.1), although in *Xiphinema* the particles also line the cuticle in more posterior parts of the alimentary canal (such as the oesophagus). In trichodorids,

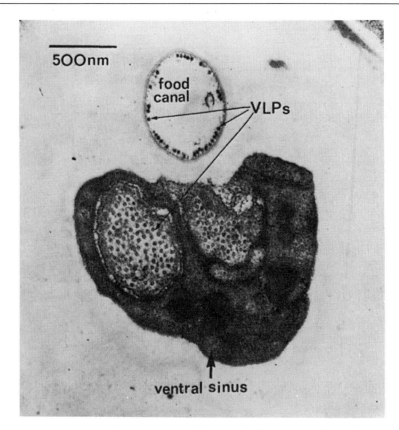

Figure 3.1 Transmission electron micrograph showing an oblique transverse section through the posterior spear/anterior odontophore region of *Xiphinema diversicaudatum* (Nematoda) that had been exposed to plants infected with strawberry latent ring spot (nepovirus). Spherical virus-like particles line the food canal of the spear and pack the ventral sinus that is an anterior extension of the oesophagus. Courtesy of W.M. Robertson and C.E. Taylor, Scottish Crop Research Institute.

the distribution of rod-shaped tobravirus-like particles (Figure 3.2) resembles that in *Xiphinema* except that the onchiostyle does not seem to adsorb particles, even when they pack the oesophageal lumen. In both longidorids and trichodorids, virus is voided in faeces which may contaminate root surfaces, thereby contributing to the viruses in soil water while really isolated from replicative opportunities.

Several orders of virus–vector specificity have been recognized. Thus, it is possible to make the generalization that nepoviruses are transmitted by longidorids, whereas tobraviruses have trichodorid vectors. Numerous viruses that are not normally inoculated to other

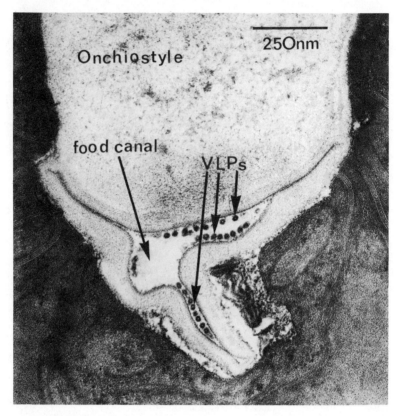

Figure 3.2 Transmission electron micrograph showing transverse section through posterior region of feeding spear (onchiostyle) of *Paratrichodorus pachydermis* (Nematoda) that had been exposed to tobravirus-infected plants. Tubular virus-like particles (VLPs) are seen 'end on' in the food canal of the stylet. Courtesy of W.M. Robertson and C.E. Taylor, Scottish Crop Research Institute.

plants by nematodes are nevertheless ingested when nematodes feed on infected plants (Jones *et al.*, 1981; Brown and Trudgill, 1983). Furthermore, some populations of species that include virus vectors are unable to transmit viruses that occur in plants in their vicinity. By contrast, tobraviruses are usually experimentally transmissible by a range of trichodorid species, yet local populations seem to vary as vectors and there are few data implying that transmission is best with a virus isolated from the same site as the vector. During a series of detailed reassessments, the earlier experiences in The Netherlands and in Scotland were confirmed and a close association was demonstrated between trichodorid species and tobravirus serotypes (Brown *et al.*,

1989; Ploeg *et al.*, 1991, 1992, 1993). At a different level, nepoviruses differ in their ability to be transmitted by, for example, *Longidorus elongatus* which transmits the serologically distinct raspberry ringspot and tomato black ring but not arabis mosaic virus. Even serologically related but distinguishable viruses differ. Thus, *Xiphinema diversicaudatum* transmits arabis mosaic virus but not another virus with which it has a few antigens in common, grape vine fan leaf virus (vectored by *X. index*). Although data are few, it appears that vector specificity lessens as the number of common antigens held by viruses increases and it seems plausible that only virus particles invested in protein coats having an appropriate surface charge are adsorbed to vector cuticle. Conceivably vector saliva changes the pH, thereby facilitating release. The critical 'mix and match' experiments addressing the molecular basis of virus–vector specificity are now possible. A substantial number of nepoviruses with different natural vectors have been sequenced *in toto* and contrasted to reveal, among other things, candidate amino acid groups (VQV in the context of *Xiphinema diversicaudatum*; Kreiah *et al.*, 1994) amenable to investigation by mutagenesis and assay either as transgenic genes in plants acceptable to vector nematodes or as biologically active transcripts for mechanical inoculation. Once vector receptor sites are recognized (if they occur) they can be avoided in the design of transgenic 'resistance' genes.

3.4 INSECTS AS VIRUS VECTORS

3.4.1 HOMOPTERA

About three-quarters of the known plant pathogenic viruses are naturally transmitted by insects, and homopteran species in four main groups are the most commonly implicated (Figure 3.3). Aphids (*Sternorrhyncha, Aphididae*) exceed other types of vectors in the number (about 150) and variety of viruses transmitted. Hoppers (*Auchenorrhyncha*) of various sorts, e.g. leafhoppers (*Cicadellidae*) and planthoppers (*Delphacidae*), have been implicated as vectors for about half as many viruses as have aphids, whereas whiteflies (*Aleyrodidae*) and mealy bugs (*Pseudococcidae*), which are most diverse and numerous in the tropics/subtropics, are lesser but nevertheless important pests and virus vectors (Carter, 1973; Bird and Maramorosch, 1978; Byrne and Bellows, 1991).

Even though aphid–virus relationships have been intensively studied, only some 300 species of aphids have been tested as virus vectors, a small part of the 4000 species presently recognized. Furthermore, no two individuals within a population of an aphid species are identical; even the progeny from a single female raised on indistin guishable leaves vary physiologically and/or morphologically.

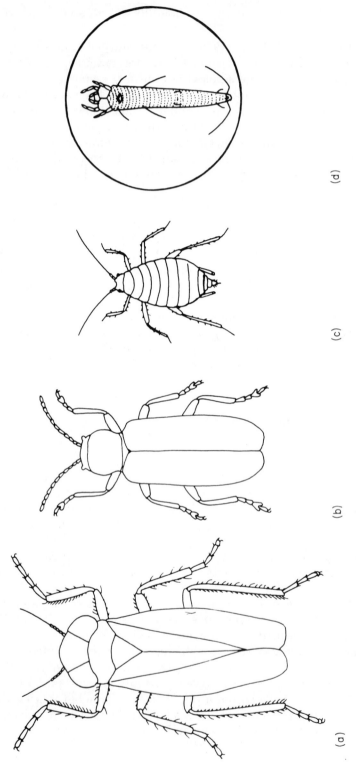

(a)

(b)

(c)

(d)

Figure 3.3 The size range of some arthropod vectors of plant viruses illustrated with four examples: (a) leafhopper (*Nephotettix* spp.) length *c.* 5 mm; (b) beetle (*Oulema melanopa*) length *c.* 4 mm; (c) aphid (*Aphis fabae*) length *c.* 2 mm; (d) eriophyid mite (length *c.* 0.2 mm) at greater magnification.

Aphids produce several short-lived generations each year with only one reproducing sexually. Typically the sexual stage gives rise to eggs that over-winter. Overcrowding and changed quality of food plants (because of age) contribute but day length in autumn provides the main stimulus of egg-laying (Figure 3.4). Plants on which aphids mate and lay eggs are described as primary hosts. The immature wingless females, which in spring emerge from eggs, moult four times before becoming adults which are also commonly wingless. Secondary hosts are those on which females in the population are capable of giving pathogenetic birth to live young (the eggs hatch just before emergence and at parturition, nymphs produced may also contain the embryo of the next generation). In this way, it is theoretically possible for a single female to produce progeny that would cover the land area of the world in 15 generations if food were unlimited and neither pests nor predators wrought havoc with the population. Fortunately, these factors impose constraints at an early stage. Nevertheless, excessive infestation may cause individual host plants to collapse with such rapidity that the wingless aphids are left marooned to crawl over the soil in a frequently futile search for a new source of nutriment. After two or three genera-

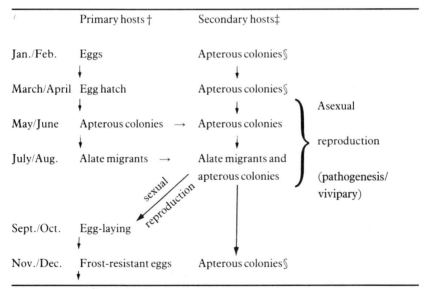

* Some aphids use only one type of herbaceous host, e.g. cabbage, throughout the year and others seem confined to woody hosts.
† *Euonymus* spp. (Spindle) Guelder rose and *Sorbus* spp. for *Aphis fabae; Prunus* spp. (Cherry) for *Myzus persicae.*
‡ Numerous herbaceous plants including crops such as sugar beets, beans and potatoes.
§ Survival over winter on secondary hosts in mild weather.

Figure 3.4 Annual life of aphids* in Britain.

tions, the apterous females may produce winged females which migrate to secondary hosts. During migration, aphids feed and breed on volunteer crops or biennials in their (second) seeding year. Additionally, polyphagous aphid species can alight and feed briefly on hosts that they would not normally prefer. Primary hosts rarely harbour viruses that naturally infect secondary hosts, but viruses of the intervening casual hosts commonly do. Winged females colonizing intermediate hosts (often crops) commonly initiate viviparous generations which increase the population; then more winged generations are produced and migrate again, thereby contributing to the movement of viruses within and between crops. Other homopterans are somewhat less fecund and opportunistic than aphids. Additionally, they differ markedly in their potential for dispersal. Thus, mealy bugs and whitefly pupae have sedentary habits, whereas hoppers are active and have life spans (varying with temperature to 4 months) that include periods of nomadic wandering. Hoppers have larval stages that jump and adults are capable of flight but most populations include morphs with only short wings, hence flying ability varies.

The best-studied whitefly vectors will feed on numerous species of natural hosts, and the handful of aphid species (e.g. *Aphis fabae, A. gossypii* and particularly *Myzus persicae*) most commonly used in experiments have a very wide host acceptance but seem exceptional in this property. Most aphid and hopper species are specific (at least in being restricted to host plants of one family). Some hoppers breed on only one genus or even on one plant species but others have complex life cycles, and alternating generations involving both woody perennial and herbaceous annual plants.

Feeding habits

Aphids feed by imbibing cell sap through stylets generally capable of penetrating cell walls in about 10 seconds, the epidermal cell layer in about a minute and the mesophyll/cortical tissues in periods that have been variously measured in minutes, hours or days (Figure 3.5). The majority of aphids insert their stylets between cells until reaching a phloem sieve tube which is delicately punctured and, when feeding is terminated, the stylets are withdrawn along the entry track. Immediate trauma is minimal; the puncture is 'repaired' and the stylet track filled with a gelatinous secretion. Aphids presented with a new leaf surface probe the cell (usually epidermal) sap, possibly to judge host suitability. Feeding/probing involves imbibition but also secretion which has a structural function (gels that repair and support the stylet track) and enzymatic activity (pectinase facilitating stylet penetration and polyphenoloxidases which might tend to damage virus particles).

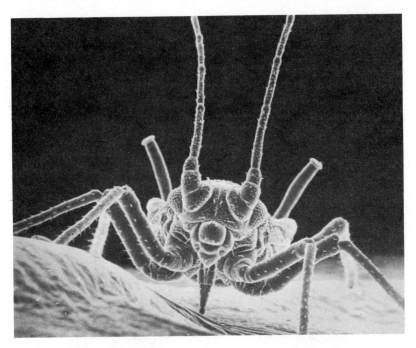

Figure 3.5 Scanning electron micrograph showing an aphid (*Myzus persicae*) on a leaf surface (×75). Courtesy of Long Ashton Research Station, University of Bristol.

Whitefly vectors and mealy bugs seemingly resemble aphids in penetrating plant tissue intercellularly to feed in phloem. However, species/genera and individuals differ in their habits, and both sexes are not equally effective vectors in all instances. Thus *Planococcoides* (*Pseudococcus*) *njalensis* is a mealy bug species with a notable predilection for rapid insertion of stylets into phloem of cocoa stems; other mealy bug species preferentially feed in leaf phloem or in other tissues. Consequently, even insects in the same family have differing abilities to acquire/transmit virus isolates, having distinct histological preferences within their plant hosts. Hoppers are particularly diverse in their feeding habits; many hopper species feed in the water-conduction xylem and imbibe large volumes of sap (Figure 3.6). Although hopper species which are notable virus vectors feed in the phloem, they tend to reach this deep-seated tissue after penetrating through, rather than between, plant cells (Nault and Ammar, 1989). Consequently, these insects cause more feeding damage to their hosts than do aphids.

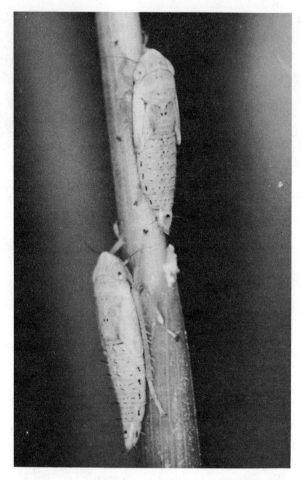

Figure 3.6 Photograph showing two leafhoppers (*Nephotettix virescens*) on a rice stem (×15). Courtesy of NERC Institute of Virology.

Virus retention and relationships with vectors

Mechanical, non-circulative association

Virus transmission is conveniently described in terms based on the ecologically important property of retention. Since 1939, the association has been classified as non-persistent or persistent, depending upon whether ability to transmit was manifest only for minutes or during weeks. Figure 3.7 illustrates four sorts of virus retention by aphids. As knowledge of the associations between viruses and their (particularly aphid) vectors increased, other more generally applicable terminology has evolved. Virus-transmission systems (and the viruses involved) are

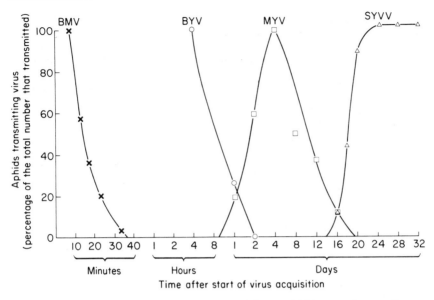

Figure 3.7 Transmission/retention records of four aphid-borne viruses. Reproduced with permission from Gibbs and Harrison (1976). Beet mosaic virus (BMV); beet yellows (BYV) and malva yellows (MYV) transmitted by *Myzus persicae* (Watson, 1946; Costa *et al.*, 1959). Sowthistle yellow vein (SYVV) multiplying in *Hyperomyzus lactucae* (Duffus, 1963).

currently described as non-circulative and circulative (see Table 3.2).

Non-circulative (including non-persistent) viruses are characterized by acquisition and inoculation after a few seconds' feeding (commonly in the range 15–60 seconds for aphids). Most of the known aphid-borne viruses (e.g. cucumoviruses, potyviruses) are non-circulative. As a rule, ability to transmit non-circulative viruses is lost by aphids within minutes of acquisition and after moulting, indicating that the viruses are mechanical contaminants. Moulting involves shedding and replacement of the external body covering with stylets plus regions of the pharynx, as well as the anterior and posterior parts of the gut lining, and knowledge that inoculativity persists after moulting is generally taken to indicate that a virus circulates in haemolymph or indeed replicates within the vector's cells. However, even among homopteran–virus combinations, in which transmission is thought to be mechanical and non-circulative, there is considerable divergence in the timing of the differing phases of acquisition/transmission. After acquisition, the time spent fasting or feeding, as well as the onset of moulting, all influence virus retention, as measured by the vector's ability to initiate disease. The host selection (probing) behaviour of homopterans serves to contaminate the anterior part of the alimentary

Table 3.2 Properties of viruses that are mechanical contaminants (non-circulative) or circulative in their arthropod vectors

	Mechanical association	Circulative association
Acquisition/transmission cycle	Short (minutes)	Long (hours–days)
Latent period between acquiring and transmitting	Not required	Required
Fasting before acquisition	Increases likelihood of transmission	Does not increase likelihood
Ability to inoculate	One or few organisms	Several organisms
Virus–vector specificity	±	+
Retention of virus after moulting	–	+
Site of inoculation	Superficial (epidermal)	Connective/vascular
Virus in haemolymph	–	+
Acquisition of ability to inoculate after experimental injection	–	+
Transmission to progeny in eggs	–	+ (if propagative)

system with virus-laden sap. Insofar as is known, virus persists either by selective adsorption to the homopteran cuticle or intimate association with components of food that themselves adhere to vector surfaces. Even though the association between viruses and homopteran vectors may be very short-lived, there is usually a measure of vector–virus specificity which is lacking in the more casual mechanical transmission of myxoma virus by dipterous flies (see p. 79) and other insects.

With few exceptions, the association between plant pathogenic viruses and their vectors is very specific; it is also liable to be complicated, as when potyviruses elicit the production by their plant hosts of proteins that facilitate virus–vector interactions (Govier *et al.*, 1977).

Circulative association

A potential for viruses to invade the haemolymph of vectors should be the justification for their description as circulative. However, this property is normally inferred from knowledge that virus transmission occurs even when moulting intervened between acquisition and inoculation feeding (Sylvester, 1980). Alternatively, or additionally, knowledge that vectors do not transmit viruses immediately after acquisition feeding is taken to be indicative. Circulative viruses include geminiviruses (in hoppers) and luteoviruses (in aphids and whiteflies), as well as rhabdoviruses and reoviruses (in aphids and in hoppers).

Geminiviruses and luteoviruses have been detected in haemolymph but do not seem to replicate and do not noticeably affect vector physiology or anatomy during the period (1–3 months) while inoculability is retained. Characteristically, the ability of vectors to inoculate plants with these viruses diminishes in the absence of renewed opportunity to feed on virus-infected plants. Insofar as is known haemolymph acts as a reservoir of virus in these circulative (non-propagative) systems of transmission. A measurable latent period is normally characteristic of circulative relationships between viruses and their vectors although the duration of the latent period varies widely, being 3–4 hours for the geminivirus associated with sugar beet curlytop disease in the hopper species *Circulifer tennellus*, 6–12 hours for the geminivirus associated with maize streak disease in hoppers of the genus *Cicadulina*. To some extent the duration of the latent period is determined by ambient temperature which affects the rate of virus replication in the plant source and hence concentration.

Individuals within populations of vector species differ in their abilities to act as vectors, and genetically controlled differences in the permeability of gut wall tissue have been identified in *Cicadulina* spp. (Rose, 1978). Indeed, blockages at this phase of acquisition seem to explain most of the variation: experimental puncturing of the gut or injection of virus directly into the haemocoele causes non-vector insects to act as vectors. Interestingly, vectors can carry more than one distinguishable virus simultaneously within the circulating haemolymph. Distinguishable virus isolates seem to be subjected to forces of diffusion rather than differential rates of sequential infection while passing from haemolymph into the salivary glands, where they are available to inoculate plants.

Propagative association

Only two groups of circulative plant pathogenic viruses (rhabdoviruses and reoviruses) are known to have a propagative relationship with (i.e. infect) their homopteran vectors. Of the 30 rhabdoviruses that have been recognized in plants and for which vectors are known, six are transmitted by leafhoppers, six by planthoppers and the residue by aphids or non-homopteran insects; the plant-infecting viruses in the Reoviridae seem to be exclusively transmitted by hoppers. The relationship between maize rough dwarf phytoreovirus and its planthopper vectors (Harpaz, 1972) is more or less typical of the propagative–circulative relationship between viruses and their homopteran vectors. After a latent period of 10–20 days, during which time virus infects cells of the insect, larvae and adults may inoculate plants to which they are given access for periods as brief as 30 minutes. When virus is acquired by first-instar larvae, ability to transmit may be retained for life (c. 50 days) and during this time daily transfer to different hosts has shown that individual hoppers may infect more than 30 plants sequentially. However, different planthopper species are known to differ as vectors of this reovirus while some instars are more and others less efficient transmitters.

3.4.2 HETEROPTERA

Although others have been tested, only two species of plant bug (Heteroptera) have been identified as virus vectors (Proeseler, 1966). Most is known about *Piesma quadratum* and its association with sugar beet leaf curl virus (Rhabdoviridae). *P. quadratum* over-winters as adults that have flown or crawled into crevices in tree bark during autumn. In spring, the piesmids progress from the forest edge to feed on weeds and then to sugar beet seedlings. In beet, the characteristic rhabdovirus particles seem only to occur in phloem parenchyma of leaves and roots. Because of this restriction, 30-minute probes are the minimum required to allow acquisition by piesmids; acquisition access periods of 1 week or longer greatly increase the probability that the month depending to some extent on the acquisition access period, has been identified. Following acquisition, piesmid nymphal instars and adults remain infective throughout their life span, but do not seem to transmit virus from females to offspring via eggs. Bugs acquire an ability to inoculate beet with the virus following injection with sap from virus-infected plants, and injected bugs have been used to show that the virus multiplies in its vector as well as in the plant host. Intriguingly these studies on multiplication showed that irrespective of whether inoculum came from plants or other insects, serial propagation in bugs enhanced pathogenicity, increasing mortality for the insects. Furthermore, infectious virus has been detected in the cells lining the alimentary canals of over-wintering bugs that had abstained from feeding for several months. When in plant sap, sugar beet leaf curl virus loses infectivity in about a week, thus propagation and persistence in piesmid vectors is important ecologically.

3.4.3 THYSANOPTERA

When the larvae (not adults) of thrips (Thysanoptera) feed on epidermal tissues of plants infected with tospoviruses, they may acquire an ability to transmit the viruses which is retained for life (20–40 days depending on the species). Tospoviruses are retained through moulting, pupation and emergence into adulthood, implying a circulative association. Furthermore, several lines of evidence might suggest

tospoviruses replicate in cells of their thrips vectors. Thrips are cosmo-
politan and many species are polyphagous, but the western flower
thrips (*Frankiniella occidentalis*) seem to be outstandingly efficient vectors
of tospoviruses. Interestingly, survival and fecundity of thrips is
diminished when they feed on plants infected by tospoviruses (German
et al., 1992): a point of contrast with aphids (Blua *et al.*, 1994). Thrips
feed on pollen and, since 1987 (Sdoodee and Teakle, 1987), thrips-
mediated transmission of ilarviruses (which had earlier been considered
to be vertically transmitted in pollen to seed) has been reported on
several occasions (Mink, 1992). In addition, thrips transmission (without pollen involvement) has been described for sowbane mosaic
sobemovirus (Hardy and Teakle, 1992) and for tobacco ringspot
nepovirus (Messieha, 1969). These are among the very few exceptions
to the 'rule' that a virus having vectors in one major taxon is not also
transmitted by organisms of another.

3.4.4 COLEOPTERA

Leaf-feeding fleabeetles (Chrysomelidae), weevils (Curculionidae) and
beetles in a few other families (e.g. Coccinellidae) acquire, retain and
transmit comoviruses, tymoviruses, bromoviruses and sobemoviruses
as well as tobacco mosaic tobamovirus. Excluding bromoviruses, for
which vectoring by nematodes, eriophyid mites or fungi has been
claimed, and the omnipresent tobacco mosaic tobamovirus, the transmissions seem to be beetle-specific (Walters, 1969). To acquire viruses,
beetle larvae and adults require access to infected plants during periods
of 1 or 2 days. Because a beetle may feed (albeit intermittently) for
several days before transmission, it is difficult to determine the period
of virus retention and the occurrence of a latent period. Retention
times seem to vary with the vector species, the ambient temperature
and acquisition access time, but the virus concentration in the source
plant is also important. Thus, cowpea mosaic comovirus is more concentrated in cowpea (*Vigna sinensis*) than in soybean (*Glycine max*), and,
when cowpea is a source, the virus is retained by the vector for the
longer period. Perhaps because of host competence, retention time has
been reported as 7 days or, at another extreme, as several months
(when the beetles over-wintered and presumably did not feed).

Inoculation of plants seems to be due to feeding, but other mechanisms, such as abrasion with claws contaminated with virus, cannot
be ruled out. Beetles are unusual in that haemolymph leaks from joints
between articulating segments (in coccinellids the exudate is yellow
and is thought to deter predators). Viruses capable of infecting plants
have been detected in this exudate, but it is presently difficult to assess
the relevance of this source of virus in beetle transmission. It has been

suggested that beetles inoculate plants because they regurgitate virus-contaminated food from an earlier meal at subsequent feeding sites, but beetles ingest numerous viruses and both those which are beetle-vectored and those which are not so vectored may be present in regurgitant. Even though leaf surfaces may be contaminated in this way experimentally (and probably this is normal in the field), it is not uncommon for closely observed experimental plants to remain healthy. Furthermore, if contamination of the mouth parts was the normal mechanism of transmission, contagious viruses (such as tobacco ringspot nepovirus which has been used experimentally) should be readily transmitted by beetles, but this has not yet been observed. For these and other reasons the simplistic view that beetle transmission occurs by mechanical contamination has been questioned (Gergerich and Scott, 1991). It has been shown that ribonuclease in beetle regurgitant selects beetle-transmissible (Gergerich *et al.*, 1986) virus by a process which seems to depend on the rate and pattern of virus escape (into vascular tissues) rather than structural resistance of virions to enzymatic attack (Field *et al.*, 1994).

Beetle-vectored viruses acquired during feeding (but not those that are not transmitted) rapidly enter the haemolymph, possibly explaining retention and specificity. Intriguingly, sparse data suggest that viruses in beetle haemolymph are more concentrated than in the sap of plants on which the animals have fed. This might imply replication in the vector but more probably reflects selective permeability of the gut wall and uneven distribution of viruses in the plant host or some other technical problem.

3.4.5 DIPTERA

Mechanical virus–vector association

Dungflies (Phoridae) are suspected to be significant virus vectors in commercial mushroom farms because they transport infected fungal spores and mycelium. Furthermore, larvae of dipterous flies that tunnel through leaf mesophyll and mosquitoes feeding on plants (which is more common than often imagined) may in rare instances facilitate virus transmission as when affecting pollination. However, even though mechanical contamination of ovipositors or other appendages with viruses such as tobacco mosaic tobamovirus or sowbane mosaic sobemovirus probably accounts for a few instances of inoculation, dipterous flies seem trivial agencies of horizontal transmission for plant pathogenic viruses.

By contrast, dipterous insects transmit in a non-circulative (mechanical) way several viruses that are concentrated in the skin and peripheral blood of vertebrates. It has indeed been proposed that viruses which have the facility to cause the clumping of blood cells (haemagglutination) block capillaries and thereby enhance their selection prospects. Myxomavirus (Poxviridae) is transmitted by a wide range of biting insects including fleas and mosquitoes and tabanid flies, which feed by excavating a depression into which blood seeps and from which meals are imbibed. These have also been implicated as non-circulative vectors of togaviruses (and more recently retroviruses). Additionally, the haematophagous Muscidae, including those that feed by sponging fluids from wounds as well as those with piercing mouth parts, probably are occasional vectors for poxviruses.

Casual non-circulative transmission of enteroviruses (Picornaviridae and Reoviridae) which are voided in faeces undoubtedly occurs via coprophagous flies, the faeces–fly–food route complementing opportunities for spread of these viruses which are more commonly transmitted in a direct faecal–oral way.

Circulative–propagative association

The importance of dipterous insects as vectors is largely attributable to the association of polyphagous mosquitoes with circulative–propagative togaviruses causing fever, encephalitis and death in mammals (McLintock, 1978).

The larvae of some mosquito species are predatory and facultatively cannibalistic and may become infected with viruses which they ingest with their food; however, male mosquitoes do not suck blood, and only the adult females seem important as virus vectors. Females require blood meals to stimulate egg development and, after a few days' rest, oviposition. Up to 20 batches of eggs may be produced during a 100-day period. The feeding apparatus of female mosquitoes consists of two fine tubular stylets surrounded by two pairs of cutting stylets which, by repeated thrusting and partial withdrawal, facilitate entry of the tips of the feeding tubules into superficial capillaries in the skin of their hosts. Fluid containing anticoagulants from the mosquito's salivary glands enters the wound down one tubular stylet while blood is actively sucked up the other. Eggs are routinely laid in water, but, although mosquito species are very selective, the number of micro-habitats available and the diversity of mosquitoes adapted to each specialized environment ensures continuity of supply (Gillett, 1971). The most important yellow fever vector, *Aedes aegypti*, is a supreme opportunist which rapidly exploits domestic rubbish containing water, tins, discarded tyres, etc.

Following inoculation by injection, several species of mosquito will support most mosquito-borne arboviruses. However, this experience is not mimicked in nature. Viruses differ markedly in their ability to infect insects that imbibe them with blood meals, and natural populations of dipterans tend to be characterized by highly specific virus–vector relationships. The time that must elapse between a blood feed and ability to inoculate, the latent (= extrinsic incubation) period, varies with the genotype of vectors, viruses and ambient temperature but is commonly 2–3 weeks. Additionally, the amount of virus in the viraemic vertebrate host (and the duration of viraemia) influences the prospect of virus acquisition and the duration of the extrinsic incubation period.

Even when simultaneously carrying and infected by two distinguishable togaviruses, mosquitoes usually seem to be unaffected and remain inoculative for life. Semliki Forest virus (Togaviridae) is, however, exceptional by causing damage to salivary gland cells of vector *Aedes aegypti* that rapidly (*c.* 10 days after acquisition from blood) lose their ability to inoculate vertebrates with this virus.

3.5 ACARINA AS VIRUS VECTORS

3.5.1 AS VECTORS FOR PLANT PATHOGENIC VIRUSES

The authenticated viruses transmitted by plant-feeding eriophyid mites (*Eriophyidae*) have flexuous rod-shaped particles about 700×15 nm and in this respect resemble the non-circulative aphid-borne potyviruses. Partly because of their small size (*c.* 250 μm in length), knowledge concerning the biology of eriophyids is sparse, but warm dry wind is thought to be the principal means of dispersal. Occasionally, insects and birds may also carry eriophyids between plants on which they can themselves crawl (very slowly). With few exceptions, eriophyids seem highly host-specific (even to clones of a plant species). The animals are maintained because they either infest perennials or annuals which grow in overlapping sequence throughout the year. Feeding is very superficial and is achieved by piercing cells of succulent tissue, such as buds, and ingesting sap (predigested with salivary excretions, which incidentally cause plants to be distorted or discoloured).

Nymphs and adult eriophyids (*Aceria tulipae*) acquire an ability to transmit wheat streak mosaic virus (WSMV) during feeds of 15 minutes duration, whereas *Abacarus hystrix* acquires ryegrass mosaic virus given access to virus-infected plants for periods of 2–12 hours. Nymphs continue to transmit after a moult and adults retain an ability to transmit for life (up to 9 days at 20–25°C) but there is no evidence that the virus is transferred to progeny via eggs. Although not known to be transmitted by eriophyids, the spherical particles of brome mosaic bromovirus and the stiff flexuous rods of barley stripe mosaic hordeivirus may be ingested with the distinctive flexuous particles of WSMV.

All three particle types accumulate largely but not exclusively within the lumen of the midgut, and because the fore- and hind-gut linings are shed at moulting, the midgut seems an important source of inoculum. Local multiplication of WSMV and brome mosaic virus cannot be excluded because virus-like particles of both types reportedly occur in eriophyid cells (Takahashi and Orlob, 1969). However, transmission of brome mosaic virus by these mites has not been recorded. Eriophyids inoculate plants with viruses such as WSMV when feeding through leaf surfaces that have been contaminated with virus in salivary/faecal excretions or by direct introduction into a feeding site by reversed flow of ingested material.

Numerous viruses have been found in an infective state in the alimentary canals of spider mites (Tetranychidae) feeding on infected plants, and these mites excrete the viruses in faeces. However, there is no confirmed report that these viruses are transmitted from one plant to others by spider mites.

3.5.2 AS VECTORS OF VIRUSES PATHOGENIC FOR VERTEBRATES

For humans, tick-borne viruses seem less important than those with mosquito vectors, though this is probably not true for wild birds or wild animals and the pattern may be changing (Hoogstraal, 1981). About 100 flaviviruses (Togaviridae) have been recognized as having a circulative-propagative vector relationship with a similar number of tick species. The vectors are in two main groups: about two-thirds are hard-bodied ticks (Ixodidae) having a variety of life styles but a somewhat insecure existence, waiting in rock crevices or on vegetation for chance to supply them with a vertebrate host (Needham and Teel, 1991); the remaining vectors are soft-bodied ticks (Argasidae) which are intermittent feeders, characteristically infesting birds' nests, rodent burrows, etc., where they can more confidently predict the delivery (and the species) of their next blood meal (Figure 3.8).

In relation to human disease, the most polyphagous vectors are the most dangerous, but the viruses that periodically affect humans are probably normally maintained by tick vectors that routinely feed on very few species of wild vertebrates. Indeed, ticks seem remarkably adapted to specific ecological niches, vertebrate host, topography and climate (Hoogstraal, in Gibbs, 1973). Ticks are resilient opportunists that rapidly engorge 10 times their body weight of host fluids (notably blood) before detaching themselves from their hosts. On migrating vertebrates, male hard-bodied ticks may remain for months and females for lesser periods, and are thereby transported great distances. By contrast, the larval stages (up to eight) of soft-bodied ticks are commonly deposited near where they hatched from eggs and the adults are

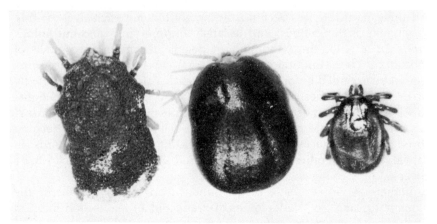

Figure 3.8 Three ticks (×6). On the left is an engorged Argasid; the others are Ixodid ticks. Courtesy of NERC, Institute of Virology and Environmental Microbiology.

rarely found on their hosts. Reflecting high mortality, individual female ixodid ticks typically produce thousands of eggs, although the soft-bodied ticks are less fecund. All mobile stages of tick vectors can acquire and inoculate viruses, with the injection of tick saliva containing virus being the most important known route of introduction in vertebrate hosts. Viruses pathogenic for vertebrates are retained after moulting and perennate in populations of ticks because adult ticks live up to 10 years. Additionally or alternatively, viruses are transmitted via eggs. As with arboviruses generally (Hardy *et al.*, 1983), virus acquisition by ticks has been assumed to require absolutely an adequate concentration of viruses in blood. However, laboratory tests with an orthomyxovirus-like-agent (Thogoto) tantalizingly showed that virus is translocated from infected ticks to others feeding on the same guinea pig yet in the absence of detected viraemia (Jones *et al.*, 1987, 1990). Although tick saliva has been implicated as a potentiating agent, neither the mechanism nor the significance in nature of such tick–tick translocation is known.

3.6 VERTICAL TRANSMISSION OF VIRUSES IN INSECTS AND TICKS

Virus transmission with eggs of the vector seems to be an important maintenance mechanism for the agent of African swine fever in *Ornithodorus* ticks, and Russian spring/summer encephalitis (Togaviridae) in *Ixodes persulcatus* ticks. Furthermore, circumstantial evidence (e.g. virus infection of male mosquitoes) suggests that mosquito-borne togaviruses may be transmitted similarly.

Among lepidopterous insects, two types of egg-associated virus transmission have been distinguished. One is transovum, when eggs are contaminated outside the ovary (e.g. during oviposition or subsequently by virus in rain splash: from soil, from decomposing larvae or from faeces). Transovarial transmission differs in being transmission from adult female to egg *in utero*. The former seems the commoner method whereby lepidopterans become infected with baculoviruses (Longworth, in Gibbs, 1973), but there is a paucity of critical data and the distinction cannot be made in all instances.

Most is known about the relationship of plant pathogenic luteoviruses, rhabdoviruses and reoviruses with eggs of their aphid or hopper vectors but many important questions remain unanswered. For example, potato yellow dwarf virus (Rhabdoviridae) has two serologically related but vector-specific forms (*Aceratogallia sanguinolenta* transmits one and *Agallia constricta* the other). Interestingly, only the *A. constricta*-vectored form of the virus seems to be transmitted to progeny hoppers via eggs. Intensive study (through seven generations that had no access to extraneous sources of infection) of a reovirus (rice dwarf) that is not known to be carried in sperm but for which up to 80% transovarial transmission is recorded, illustrated one interesting way in which eggs of a leafhopper vector became associated with virus.

The existence of an intimate relationship between many insect species and a diverse array of antibiotic-sensitive intracellular microorganisms (assumed symbionts) has been recognized for many years. In homopterans, the associations are particularly close, and elaborate methods ensure transovarial transmission of the non-motile 'symbionts'. Interestingly, electron microscopy revealed virus-like particles adhering to the surface and/or within the cytoplasm of the membrane-bound 'symbionts'. Although rice dwarf reovirus seems to infect vector progeny *in utero* when virus-contaminated 'symbionts' from the parent enter oocytes as yolk is being formed, there is no direct evidence that the virus replicates in the 'symbionts'. Proteins produced by the endosymbionts in aphids stabilize luteoviruses and thereby facilitate their transmission (van den Heuvel *et al.*, 1994). Within a week, most progeny that acquire virus *in utero* are capable of infecting plants on which they feed but reoviruses such as rice dwarf probably do not persist indefinitely in populations of all leafhopper species that can act as vectors. In one experiment rice dwarf reovirus disappeared after only three generations, but another reovirus, associated with rice stripe disease, was reportedly transmitted through 20 generations without decline in inoculativity.

The epidemiological importance of vertical transmission in insects is uncertain because few measurements have been made on the effects of infection on vector (host) survival and fertility (fecundity). Furthermore, it is difficult to relate the proportion of infected progeny to the total reproductive potential of the insects. However, Sylvester (1969) accumulated a comprehensive body of biological information which was analysed statistically and showed that the rhabdovirus sowthistle yellow-vein virus depends on the plant components of the community for maintenance, the transovarial transmission cycle in aphids being incidental but probably not irrelevant. Intriguingly, progeny of some vectors of plant pathogenic viruses (e.g. *Inazuma dorsali*) are rarely infected transovarially, and those that are seem to die prematurely. Similarly, maize rough dwarf virus (Reoviridae) seems harmful to larval but not adult *Laodelphax striatellus* (Harpaz, 1972).

3.7 THE RANGE OF INSECT-PATHOGENIC VIRUSES

A lot is known about viruses that are pathogenic for insects, but it must be emphasized that the data collated in Table 3.3 very probably represent only the 'tip of an iceberg'.

David (1975) listed 721 insect species in which virus-like diseases were recorded, although this included some experimental transmissions. However, by no means all of the presumed viruses have been isolated and many of those that have been recognized recently are not grossly pathogenic. It is now possible to grow some viruses isolated from insects in cells cultured *in vitro*, thereby facilitating cloning as well as assay. Most viruses described in association with insect disease have, however, been maintained in whole insects, their presence being inferred from uncritical criteria such as the death of their host. Many of the described viruses were studied only after passage through numerous hosts, and distribution to different laboratories. Regrettably, the possibility that a virus isolated was not that which was earlier contracted naturally or inoculated experimentally but one that pre-existed in that host has often been ignored. Largely because numerous viruses without marked effect on host populations have been recognized during the last 20 years, this possibility is now considered. Speculative studies, using termite extracts to inoculate *Porotermes adamsoni*, *Drosphila* extracts to inoculate *Drosophila* spp., kelp flies (Diptera) to inoculate *Galleria* (Lepidoptera) larvae and bee extracts to inoculate *Apis* spp., have revealed a multitude of unsuspected viruses with RNA genomes (Gibbs *et al.*, 1970; Bailey, 1981). A few viruses isolated from insects have properties in common with viruses pathogenic for other animals or plants but some of the most studied are distinguishable because their virus particles are occluded within crystalline cell inclusions. Four groups of occluded viruses are currently recognized from insects on the basis of virus shape, the number per inclusion and the main site of accumulation in their host.

Table 3.3 The range of insect-pathogenic viruses

Categories of agent	Known natural arthropod host range, order	Similarities to viruses	
		Vertebrates	Plants
Baculoviruses	Lepidoptera, Hymenoptera, Diptera, Crustacea, Coleoptera, Trichoptera?	−	−
Cytoplasmic polyhedrosis viruses	Lepidoptera, Hymenoptera, Diptera, Crustacea, Coleoptera	+ (reoviruses; numerous homoeo-therms)	+ (Fungi and higher plants)
Entomopoxviruses	Lepidoptera, Orthoptera, Coleoptera, Diptera	+ (poxviruses; numerous birds and animals)	−
Iridoviruses	Lepidoptera, Orthoptera, Coleoptera, Diptera, Ephemeroptera, Hemiptera, Nematoda	+ (e.g. frogs, fish and swine)	± (Fungi and algae?)
Parvoviruses	Lepidoptera	+ (e.g. rabbit, monkey)	−
Rhabdoviruses	Diptera	+ (e.g. fish and numerous homoeo-therms such as bats)	+
Nodaviruses	Lepidoptera, Orthoptera, Hymenoptera, Homoptera Coleoptera,	(+)	(+)
Polydnaviruses	Hymenoptera	+	

1. Nuclear polyhedrosis viruses (NPVs) are baculoviruses which incite the production in cell nuclei of polyhedral inclusions containing numerous rod-shaped virus particles. Martignoni and Iwai (1975) recorded 337 baculovirus isolations from insects, 280 from lepidopteran hosts.
2. Granulosis viruses (GVs) differ in producing ellipsoidal inclusions

containing one or occasionally two virus particles and have been reported only in Lepidoptera. The baculoviruses (Baculoviridae) including NPVs and GVs seem to have few natural hosts and appear unique to arthropods. All the polyhedral proteins of baculovirus from Lepidoptera have serological properties or amino acid sequences in common, and polyhedral proteins from baculoviruses isolated from Hymenoptera and Diptera also have a few properties in common. On sparse evidence, Rohrmann *et al.* (1981) speculated that Diptera and Hymenoptera (orders in existence *c.* 200×10^6 years ago) were first infected with NPVs. The Lepidoptera are a more recently evolved order that has undergone extensive speciation only during the past 50×10^6 years but it is plausible that the appearance of Lepidoptera provided NPVs with a new environment facilitating their evolution. GVs appeared to have diverged from NPVs before the divergence of baculoviruses in Hymenoptera. For this reason baculoviruses are the favoured candidates for biological control; each of the other categories of viruses isolated from insects has morphological and/or chemical properties in common with viruses pathogenic for plants, poikilothermic or homoeothermic vertebrates. However, putative baculoviruses have been observed in an insect pathogenic fungus, spiders and crabs and it would be premature to conclude yet that baculoviruses will not be found in other sorts of hosts. An NPV from *Autographa californica* which has been intensively studied (Ayres *et al.*, 1994) because it has been genetically engineered to optimize effectiveness as a biopesticide may be exceptional but reportedly infects 28 insect species, seven lepidopteran cell lines, a reptilian cell line and a mammalian cell line (McIntosh and Shamy, 1980).

3. Cytoplasmic polyhedrosis viruses (CPVs) which have dsRNA genomes and icosahedral particles resembling those of reoviruses are characterized by their association with crystalline inclusions which accumulate in the perinuclear cytoplasm of host cells, largely but not exclusively in the midgut. In one report it was calculated that 10 000 CPV particles were occluded within one polyhedron. About 150 records of CPV isolation exist but it is noteworthy (although probably by no means unique to this category of virus) that, when 33 CPV isolates were compared, only 11 types were distinguished (Payne and Rivers, 1976). CPV polyhedra from different hosts have different sizes and shapes, being large and polyhedral in *Bombyx mori*, spherical in *Danaus plexippus* and small in mosquitoes (*Aedes taeniorhynchus*). Although CPVs are a heterogeneous collection, few cross-infection tests have been reported and it is not clear whether the form of the inclusion is host- or virus-directed. Understanding of the biology of these viruses is hampered by the prevalence of inapparent CPV infections.

4. Entomopox viruses are occluded within crystalline proteinaceous inclusion bodies analogous to those of NPVs. In dimensions, these poxviruses seem to be a heterogeneous group: the virus particles detected in Coleoptera are larger and those from Diptera smaller than those isolated from lepidopteran or orthopteran hosts. Interestingly, fowlpox (but not all poxviruses from vertebrate hosts) is characterized by inclusions that contain mature infective virus particles. Although there are enzymes held in common between invertebrate and vertebrate poxviruses, there are crystallographic differences between the fowlpox inclusion material and that from, for example, polyhedra of NPV from *Bombyx mori*, and few, if any, of the antigens possessed by vertebrate poxviruses are held in common with entomopox viruses (Bergion and Dales, 1971). Furthermore, unlike avian poxviruses, entomopox viruses that have been tested did not induce pocks on the chorioallantoic membranes of fertile hens' eggs

The non-occluded viruses detected in insects are undoubtedly very diverse. Thus a few virus isolates from Orthoptera have particles that are not occluded: a virus from a Malaysian population of palm rhinoceros beetles, *Oryctes rhinoceros* (Coleoptera), has been a notable model in epidemiological studies that will be mentioned later (see p. 89 and 93).

Some 30 iridoviruses, which can be divided into two main groups on the basis of their modal diameters, have been associated with a blue–green iridescent colour in the tissues of their insect hosts. The opalescence is due to the immense numbers of virus particles contained within the cell cytoplasm. Despite a paucity of critical data, Chapman (1974) suggested that all iridescent viruses with diameters greater than 170 nm are a single virus with a large host range. This assertion was premature; there is real diversity in their properties (Williams and Cory, 1994). In any event, there is very little (no) definite evidence showing that individual iridescent viruses infect more than one host under natural conditions. Nevertheless, iridescent virus isolates from a dipteran, a coleopteran and a lepidopteran were each experimentally cross-transmissible to host in these orders, and the viruses extracted from at least two of the experimental hosts were serologically indistinguishable from the inoculated isolates (Oliviera and Ponsen, 1966). No complete study has been made of the serological interrelationships of iridoviruses, and 20 years ago the situation was recognized to be confused (Kelly and Robertson, 1973). However, it is noteworthy that DNA from an iridovirus which infects amphibians has little, if any, sequence homology with DNA from three insect iridescent viruses despite gross morphological and chemical similarities of the virions.

Of the non-occluded insect-associated viruses having spherical particles, a few contain DNA and resemble parvoviruses in their chemical composition and structure (Tinsley and Longworth, 1973) but not serologically (Hoggan, 1971). The best known of these, densonucleosis virus, was described in association with a fatal disease of larval *Galleria mellonella* (Lepidoptera); there is an oft-quoted but unconfirmed report (Kurstak *et al.*, 1969) that the virus transformed mouse cells in culture, i.e. made them cancer-like.

In a review of isometric viruses from vertebrates, Longworth (in Gibbs, 1973) tentatively recognized eight groups of isolates from fruit flies (*Drosophila*) in particular but also from mosquitoes, moths, crickets and beetles. A common feature of these viruses is their bipartite RNA genomes and their superficial similarity to plant pathogenic diantho-viruses. More recently, the distinctness of these agents has been recognized in the establishment of a family: Nodaviridae. One of the most intriguing nodaviruses was originally isolated from *Culex tritae-niorhynchus* mosquitoes in the village of Nodamura, Japan, after which the isolate, the family and the genus was named. The Nodamura virus multiplies in other culicine mosquitoes and in *Ornithodoros* ticks without killing them but kills moth larvae (*Plodia interpunctella*), bees and suckling mice. Nodaviruses are characterized by their promiscuity, stability, fecundity and ability to establish latent infections in a diverse range of invertebrates. A serologically unrelated virus from beetles (*Heteronychus arator*) is not known to be so catholic in its experimental host range and is distinguishable from the *Culex* (Nodamura) isolate in not replicating in baby hamster kidney or mouse cells or in cultured cells of four invertebrates. These two viruses are distinguished from plant pathogenic viruses with bipartite genomes in additional properties including the protein and nucleic acid molecular weights. However, some nodaviruses isolated from insects replicate in plants and one probable nodavirus (striped jack nervous necrosis) has been isolated from a fish in which virus-like particles occurred in the brain and were associated with abnormal behaviour and economic losses in crowded hatcheries.

The other RNA-containing viruses from insects form a heterogeneous collection which are presently too little studied to justify generalization. Nevertheless, many unexplained but interesting phenomena involving these viruses have been reported and a suite of serologically related isolates, including Cricket paralysis virus and *Drosophila* C virus, having properties in common with enteroviruses (Picornaviridae) deserves special mention because of the ecological implications of the available data.

Cricket paralysis virus, although originally isolated from Australian field crickets (*Teleogryllus oceanicus*), can replicate in more than 35 different insect species distributed between five orders. Indeed, viruses which react with serum prepared against Cricket paralysis virus have been found in natural populations of lepidopterans and orthopterans in Australia and Sarawak. Additionally, the distinguishable *Drosophila* C virus has been detected in 28 of 163 laboratory populations of *D. melanogaster* from North Africa and the Caribbean. Despite their apparent ubiquity, these virus serotypes have not been associated with natural epizootics.

3.8 CONSEQUENCES OF VIRUS INTRODUCTION INTO INVERTEBRATE POPULATIONS

Following movement, in 1955, of *Pieris brassicae* (cabbage white butter-flies) from mainland Europe, where GV was present, to the UK where the virus was then unknown, wild *P. brassicae* were virtually eliminated and similar experiences with *P. rapae* in North America leave little doubt about the potential of GVs for limiting host insect populations (David, 1978). Deliberate introduction of diseased insects, though unwise when the associated viruses are unknown, has been used to dramatic effect in the islands of the Indian Ocean and Oceania. An introduced baculovirus without known contamination with different viruses significantly reduced populations of the palm rhinoceros beetle (*Oryctes rhinoceros*) in Fiji, Mauritius, Tonga, Wallis Island and Western Samoa (Entwistle *et al.*, 1983). Similarly, the picornavirus from *Gonometa podocarpi*, pests of *Pinus patula* in Uganda, and the distinct agents (Tetraviridae) from *Darna trima* (pests of oil palms in Sarawak) proved spectacularly effective in killing these insects when man deliberately introduced them into pest-infested regions where they were not earlier known to occur.

However, viruses have not often been shown to cause changes in population density which are commonplace among invertebrates in nature. Effects attributable to insect pathogenic viruses can be considerable but most obvious when humans have provided a monocrop environment facilitating the build-up of intense host density. Forestry is notorious in this respect and has provided several systems which have been used to assess prospects of using viruses as agents of biological control.

One such model is the European spruce sawfly, *Gilpinia hercyniae* (Hymenoptera), which became established in North America and flourished in the absence of natural enemies, so that 2000 square miles of spruce in Canada were infested when the pest was first recorded in 1930 (and an additional 10 000 square miles of infested trees were recognized in 1938).

About this time, a nuclear polyhedrosis disease became apparent. Probably the causal agent was unintentionally introduced (from mainland Europe where *Gilpinia hercyniae* is endemic) with its population of natural enemies then being assessed as possible agents of biological

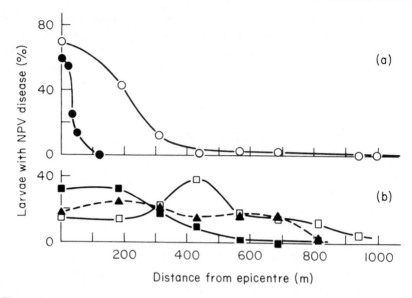

Figure 3.9 Pattern of spread of *Gilpinia hercyniae* (nuclear polyhedrosis virus) disease in Ontario, Canada, (Bird and Burk, 1961). (a) ●, September 1950; ○, September 1951. (b) ■, July 1952; □, September 1952; ▲, September 1953. The developmental sequence agrees both in form and scale with data published by Young (1974) on the spread of a baculovirus-associated disease of *Oryctes rhinoceros* in one of the Tonga Islands and with data published by Entwistle *et al.* (1983) on the spatial expansion of a nuclear polyhedrosis affecting *G. hercyniae* in Wales.

control. During the war years, the NPV spread throughout most of *G. hercyniae's* range, and thereafter, a combination of the disease and parasites/predators has maintained *G. hercyniae* as a minor rather than a major defoliating pest in North America (Entwistle *et al.*, 1983).

In 1969, *G. hercyniae* suddenly infested spruce plantations in the Welsh mountains. Considerable local density-dependent mortality associated with an NPV was recognized 2 years later, and subsequent observations have shown a rapid spread of the disease agents with significant diminution of the sawfly populations (Figure 3.9). The sources of the Welsh sawfly populations and the associated NPV are unknown.

3.8.1 REACTIONS TO INFECTION

There have been few studies on the innate virus-resistance mechanism of invertebrates but it is known that different populations of an insect species differ in their liability to die following infection with, for example, cytoplasmic polyhedrosis reoviruses. Heritable natural variation in sensitivity to viruses has been noted in a variety of insects but the operation of selection pressure is most obvious on populations of domesticated insects such as silkworms; 1000-fold differences in morbidity between geographically isolated 'landraces' of *Bombyx mori* (silkworm) have been recorded following experimental inoculations. Natural selection in outbreeding 'wild' insect populations has also been recorded (e.g. in a moth, *Eucosma griseana*, infected with GV), but the consequence of natural epizootics tends to be the removal of sensitive infectibles. Insects respond to virus infections by producing preformed and induced humoral and also cellular 'cleansing agents' (Dunn, 1986; Boman and Hultmark, 1987). Among these are interferon-like substances (Kalmakoff *et al.*, 1977) and compounds capable of agglutinating virus particles and inactivating baculoviruses (Hayashiya, 1978).

3.8.2 PERENNATION OF INOCULA

The means of perennation and the long-term effects of 'small' RNA viruses are only now being studied, but there is a considerable and expanding knowledge of NPV ecology. Once established, three factors influence the association of the viruses with their insect hosts: the stability and retention of infectivity, host and inoculum dispersal, and pathogenicity (virulence) of the viruses for their hosts (Evans and Harrap, 1982).

Outside their hosts, non-occluded viruses seem much less stable than occluded viruses but there are no critical comparative data. The proteinaceous polyhedra in which for example NPV particles are occluded, resist acid and alkali treatment over a range of pH 2–9, and NPV from *B. mori* was still infective after 21 years (but not after 37 years' laboratory storage); an NPV from *G. hercyniae* was infective after 11 years' storage at 4.5°C. These *in vitro* experiences may, of course, be exceptional. Thus, *G. hercyniae* NPV (which accumulates on leaves during late summer/autumn) is virtually completely non-infective by the following summer, although it should be remembered that cadavers of infected larvae are not all releasing polyhedra at the same time and probably provide small amounts of inoculum throughout the winter. Once host body fluids have dried on to the foliage, the associated polyhedra are not detached readily. Laboratory tests with *P. brassicae* GV showed that crude extracts of cadavers were not significantly removed during 4 months in autumn/winter or by 5 hours exposure to simulated rain or to light scrubbing with detergent following by rinsing in water. During the Northern summer and winter, physical loss of *G. hercyniae* polyhedra was exponential, the half-life on Norway spruce in summer and winter being different despite the heavier rainfall. Plausibly, foliar surfaces of different plant species differ in their relationships with deposited virus particles or polyhedra, leaf anatomy and surface chemistry playing some part in determining physical retention on the one hand and structural integrity as in alkaline dew (Andrews and Sikorowski, 1973) on the other. The number and distribution of stomata may also be important; the GV of the potato moth (*P. operculella*) enters stomata, facilitating infection of larvae mining their way through cells between the hypodermis and epidermis.

In situ inactivation, rather than decontamination, seems to explain most infectivity; loss of occluded virus (with sunlight) is considered to be particularly important. On upper leaf surfaces of cabbages leaves, *P. brassicae* GV was totally inactivated during 19 hours, whereas infectivity of agents on the undersurfaces of leaves was retained longer (Jaques, 1972). Because repeated freezing and thawing of NPVs in suspension has little effect on infectivity, low winter temperatures are probably unimportant natural inactivators. However, high summer temperatures (43°C in air; 50°C in soil) are thought to have important implications for NPV infectivity in North America (Jaques, 1975). Irrespective of any inactivating effect, ambient temperature also influences host mobility and feeding and rate of increase in host death, thereby controlling the pool of inoculum on the one hand and the size of the host population on the other.

3.9 VIRUSES IN SOIL AND PREDATORS

A large part of the polyhedra produced during natural epizootics reaches the soil and accumulates in the upper layers adsorbed to colloidal matrices. All the available evidence shows that occluded baculoviruses are neither rapidly inactivated nor physically disrupted and the soil is a long-term store of these viruses. Laboratory studies showed that *P. brassicae* GV remained infective for 2 years and little vertical movement was observed even when 48 inches (120 cm) equivalent of rain was passed through columns containing occluded GV in sand or soil.

In nature, NPVs are similarly extremely stable (e.g. an NPV from *Trichoplusia ni* retained infectivity during 4 years) and most concentrated in the top few centimetres of undistubed soils (Jaques, 1967). In forests, there are probably few ways in which this stored inoculum can reach foliage and it is largely unavailable for re-infection, but the activities of beetles, predatory invertebrates, sawfly adults (Neilson and Elgee, 1968) and small mammals may provide occasional opportunities for re-establishment on leaves. Infectible adults are liable to be contaminated with virus encountered during and after emergence, but this seems unimportant. In an agricultural context, by contrast, soil disturbance is more frequent, at least when the crops are small. Furthermore, rain-splash or wind-blow facilitate contamination of leaves on which infectible larvae are likely to feed, and irrigation by controlled flooding could also play a part in washing polyhedra out of soil and on to foliage.

Although these agencies facilitate persistence of occluded viruses at a site in a coniferous forest, polyhedral contamination of new foliage agents is involved. Birds that feed on cadavers or earthworms contaminated with NPV undoubtedly contribute inoculum; their faeces often contain viable polyhedra and the voided mass adheres to foliage and is retained for a few weeks (Entwistle *et al.*, 1977).

In Wales, 15 species of birds trapped in spruce forests during September were carrying NPV, probably that of *G. hercyniae*, and it is significant that faeces containing infective virus could be obtained in winter and spring when living sawfly larvae were absent. Birds must ingest massive quantities of occluded and non-occluded viruses in this way. Additionally, birds, in common with other insectivorous animals and their commensal mammals, are liable to inhale virus-laden air. The possibility that the vertebrate animals acquire allergies or immunity as a consequence seems worth study. Host larvae may themselves spread inoculum in faeces because some ingested polyhedra pass with partially digested food through the alimentary canal of dying larvae. Indeed, this process seems to be the most important way in which the non-occluded baculoviruses of *Oryctes* spp. is dispersed locally. Storage of this *Oryctes* virus in sawdust for 1 week reduced its infectivity to less than 1% of original, and after a month infectivity was lost. Consequently, most of the virus that may be present in the soil surrounding the eggs is inactivated during the 12-day incubation period. The adult rhinoceros beetles which are gregarious in decaying palm trunks at mating time, may become contaminated with virus, but this seems to contribute little to the infection of larvae or to adults. However, adults are readily infected at mating presumably *per os* since adults chew large cavities in their mating substratum and this is liable to be contaminated by voided virus. Wind and insects including parasites and predators undoubtedly facilitate the spread of occluded viruses, complementing the chances of dispersal provided by migrating (winged) adults.

Viruses which infect invertebrates are also transmissible by invertebrates – notably on the ovipositors of parasitic hymenopterans that lay eggs in and have larvae that develop inside host larvae. Stoltz and Vinson (1979) first described baculovirus-like particles in the ovipositors of Braconidae and Ichnemonidae. More recent work is revealing a fascinating series of interactions and properties possessed by these intimately co-evolved agents (currently Polydnaviridae) which have co-evolved with their hosts/vectors (Fleming, 1992). Less dramatically, the GV of *P. brassicae* is reportedly liable to be transmitted mechanically by stings of a hymenopterous parasite *Apanteles glomeratus*, which could acquire inoculum either by contamination from the environment or while stinging infected larvae. Furthermore, the parvovirus-like den-sonucleosis virus from *Galleria mellonella* has been transmitted from an infected to a healthy larva by the ovipositor of the parasite *Nemeritis canescens*.

Persistence of inoculum outside the host is important but it should be remembered that many viruses are able to persist for most, if not all, of the reproductive lives of their host insects. Although the NPV of *Gilpinia hercyniae* usually kills larvae within 5–10 days following inges-tion, larvae that acquired virus within 3 days of the time when the last instar stops feeding commonly survive to the prepupal over-wintering form which can survive six seasons. Conceivably, virus replicates during this time and may thereby facilitate prolonged virus survival during periods of low host density.

The dispersal habits of adult insects that are also normally chronic carriers, yet are not killed by doses of virus that are rapidly fatal for larvae, obviously play an important part in determining the pattern and rate of inoculum spread. Gregarious or weak flying hosts tend to retain inoculum locally. On the other hand, strong fliers, such as *Gilpinia*, which may fly a mile or more, facilitate jump spread (Figure 3.10). Furthermore, egg-laying habits of the host influence the number of inoculum foci; *Neodiprion sertifer* lays eggs in groups whereas *G. hercyniae* tends to lay eggs singly and thereby maximizes prospects for dispersal, not only of its offspring but also of associated viruses.

3.10 OCCURRENCE OF NATURAL ANTIBODIES TO INVERTEBRATE VIRUSES IN ANIMALS

Although numerous samples of human serum have been tested, anti-bodies against occluded viruses or the associated polyhedral proteins have been recorded only twice. By contrast, antibodies against a diverse array of nodaviruses and other picorna-like viruses pathogenic for insects have been detected on several occasions.

The morphological/chemical affinities of a virus isolated from *Gonometa podacarpi* to mammalian enteroviruses led Longworth *et al.* (1973) to test the *Gonometa* virus against sera prepared in rabbits, against a virus from *Gonometa* spp. and foot-and-mouth disease virus antigens. However, convalescent serum from pigs infected with a porcine enterovirus contained antibodies that were specific for the serologically unrelated *Gonometa* virus, as well as those against the porcine enterovirus. The antigenically distinct invertebrate virus (from *Darna trima*) also reacted with the convalescent pig sera. More signi-ficant, unselected sera from cattle (10/10), sheep (4/6), horses (4/6), dogs (1/6) and deer (10/10) also contained antibody specific for the

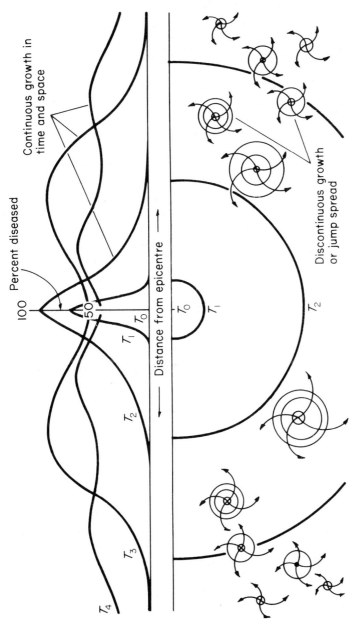

Figure 3.10 Diagram showing the spatial and temporal T_0–T_4 patterns of dispersal from a localized focus (epicentre) for pathogens indicated by the occurrence and incidence of disease. There are two components attributable to continuous and discontinuous (jump) spread of the pathogens. After Entwistle *et al.* (1983).

Gonometa virus which was isolated from an insect indigenous to East Africa and is unknown elsewhere. Several small RNA viruses possessing few, if any, of their antigenic determinants not held in common have been isolated from many parts of the world (e.g. *Drosophila* C virus in fruit flies in Morocco, France and the Antilles). A plausible explanation for the prevalence of antibodies in animals is that they were produced in response to unknown viruses (present in the UK) having antigenic properties in common with the *Gonometa* virus. Some support for this hypothesis has come from observations reported by Longworth (1978). A virus isolated from *Gonometa* is serologically related to a *Drosophila* virus, itself having antigenic determinants in common with a virus from honey bees, bee virus X.

Nodamura virus is another virus against which naturally occurring antibodies have been detected. Indeed, because such high levels of antibody against the *Culex* isolate were detected in pigs and wild birds in Japan, it was at one time thought that pigs rather than mosquitoes were the prime source (Scherer *et al.*, 1968).

Even more intriguingly, F.O. MacCallum (unpublished data) observed that some unselected sheep sera reacted with sera produced in rabbits immunized against a virus (Tetraviridae) from *Nudaurelia* spp., i.e. the sheep sera contained the antigens but were also found to have antibodies that precipitated the invertebrate virus against which they were earlier tested. This is one of very few instances implying circulation (in mammals) of an 'insect' virus eliciting an antibody response in the vertebrate. The cause or causes of these phenomena are unknown, but it is noteworthy that three sera from human workers with invertebrate viruses contained antibodies against agents to which they were exposed (cricket paralysis or *Gonometa* viruses). Furthermore, circumstantial evidence suggests that in New Zealand, cattle feeding in pasture infested with crickets infected with cricket paralysis virus had circulating antibodies against that virus (or one of its numerous serotypes). Plausibly, inhalation of virus or ingestion of cadavers elicited antibody production, although the possibility of transient infection in the cattle cannot readily be discounted. Detailed immunochemistry has been done on some of the systems mentioned above and the balance of evidence (direct and circumstantial) suggests that the serological reactions generally involved the IgM component of the globulin fraction of blood. Because IgM is normally associated with persistent antigenic stimuli of small magnitude, this might imply that the viruses did not replicate to any great extent within the animals that were bled.

Recognizing that there are thought to be about 3 million species of insects worldwide, about 10 times more than all the other animals together, and that in an acre of British farmland, numbers of individual insects are measured in hundreds of millions with a similar number of arachnids, the natural population of invertebrates is well placed to provide the presumed stimuli.

3.11 VIRUSES INFECTING OTHER INVERTEBRATES

Excluding insects, virus-like particles have been reported in amoeba and other protozoa, annelids, platyhelminths, nematodes, molluscs, various arthropods including crustaceans and scorpions, and a sponge (Diamond and Mattern, 1976; Davidson, 1981; Erdos, 1981; Wang and Wang, 1991). In some instances, attention has been drawn to similarities (usually morphological) between these arthropod-associated agents and, for example, adenoviruses, reoviruses, baculoviruses, herpes viruses, etc. However, with few exceptions, attempts at experimental transmission and cultivation of the presumptive viruses have not been successful, even in those rare instances where it has been attempted. Furthermore, the infectivity of viruses that have been cultivated has very rarely been assessed for unaffected members of the recorded host. Where detailed tests have been done, as with viruses isolated from marine molluscs, the existence of a group of agents having spherical particles *c.* 55–60 nm in diameter has been recognized to be widely distributed in limpets, clams, oysters and winkles around the shores of Britain even when geographically isolated, such as 'wild' oysters near the Isle of Mull. Interestingly, the molluscan-associated virus-like agents replicate in cells of fish grown *in vitro* and have chemical as well as morphological properties in common with the infectious pancreatic necrosis birnavirus of fish and the infectious bursal disease agents of waterfowl. It is predictable that intensive farming of marine resources will facilitate virus spread and subject diverse populations of invertebrates to stimuli that may predispose them to disease, cf. influence of toxic polychlorinated biphenyls on the incidence of virus disease in pink shrimps (*Penaeus duorarum*; Couch and Courtney, 1977).

Viruses and the terrestrial environment

4

4.1 THE RANGE OF VIRUSES IN TERRESTRIAL VERTEBRATES

Few of the viruses associated with reptiles or amphibians (Lunger and Clark, 1978) have been characterized and the list of viruses naturally infecting wild mammals (Beran, 1981; Davies *et al.*, 1981) or birds (Davies *et al.*, 1971) is very incomplete. More is known about a handful of viruses associated with economic loss in domesticated livestock (Gibbs, 1981) but, even with these, data on occurrence are liable to be affected by under-reporting because of trade implications. Although the human population in the developed world is closely monitored virologically and the occurrence of a select group of viruses in individuals is recorded, no realistic figure can yet be given for the number liable to infect humans. Estimates of virus prevalence in human populations, normally determined because of frank disease or inferred from retrospective serological surveys, are collated by the US Department of Health and Human Services, Centers for Disease Control and Prevention (Morbidity and Mortality Weekly Reports) and periodically re-assessed by the World Health Organisation, sometimes with the Food and Agricultural Organisation of the United Nations (FAO, 1977). Techniques developed to facilitate study of viruses infecting human populations are constantly revealing unsuspected virus-like agents in laboratory animals (notably rodents and primates) as well as *in vitro* cell cultures derived from these animals. Few of these agents are implicated with disease, and an increasing number (notably viruses assigned to the Retroviridae, Herpesviridae, Adenoviridae or Papovaviridae) are so closely integrated within the genetic constitution of their host cells that their distinctness is difficult to prove. Thus, all apparently normal cells of birds and rodents that have been examined in sufficient detail contain the complete genetic determinants for at least one retrovirus.

In parallel, retroviral genomes tend to include host-derived sequences and virus-associated altering of cell 'control' functions is one of several ways in which growth patterns of cells change 'naturally'. As well as hyperexpressing 'oncogenes' which have been incorporated, retroviruses elicit cell overgrowth because of their ability to insert (and excise) randomly their genomes into those of their hosts. Furthermore, during multiple cycles of replication, reverse transcription errors tend to confer growth advantages to infected cell lineages, resulting in leukaemias and solid tumours.

4.2 THE RANGE OF VIRUSES IN TERRESTRIAL PLANTS

Viruses are widely distributed in the plant kingdom and have been recorded in algae, fungi (Hollings, 1978), mosses and ferns as well as in flowering plants (Smith, 1972; Brunt et al., 1990). Furthermore, endophytic microorganisms in the vascular systems of plants are prone to be lysogenic for viruses which have been implicated with hypovirulence of spiroplasma-associated disease (Renaudin and Bove, 1994). The requirement for costly and sophisticated equipment to identify the viruses has resulted in disproportionate knowledge. Priority has been given to the study of viruses that affect food plants, and few systematic studies other than on crops have been made. However, virus infection in wild plants is indicated by four series of observations. In the largest survey, MacClement and Richards (1956) recorded an average infection of 10% after testing a total of 2193 plants in six American communities at fortnightly intervals during 4 years. Hammond (1981) found that 92 of 144 *Plantago* spp. (plantains) collected at random were affected by at least one of eight distinguishable viruses and surveys of multiple populations on a similar scale revealed up to 66% infection by viruses including cherry leaf roll nepovirus (that is seed- and pollen-transmitted) in *Betula* (birch; Cooper et al., 1984) or an aphid-borne luteovirus in *Anthoxanthum* (sweet vernal grass; Kelley, 1993).

4.3 FACTORS AFFECTING THE OCCURRENCE OF VIRUSES

4.3.1 GEOGRAPHICAL DISTRIBUTION

Maps showing the distribution of viruses and vectors are snap-shots of a dynamic process, and their interpretation is difficult because the information reflects only the thoroughness of specialists having the facilities, opportunity and inclination to investigate. Often knowledge concerns diseases liable to be caused by more than one virus, separately or indeed in combination with others. Nevertheless, all known viruses and vectors have not been found everywhere that they have been diligently sought. The reasons are imperfectly known and there is a paucity of detailed information concerning wild plants and animals, but it seems that the islands of New Zealand have few if any indigenous viruses that have so far infected crops. All (about 70) viruses were found infecting plants introduced into New Zealand and each had numerous properties in common with viruses earlier recognized in Europe or America (Matthews, 1981).

In seeking explanations, one cannot exclude the possibility that the large indigenous flora of New Zealand was infected with viruses common elsewhere in the world or that the uniqueness of the native plants is a barrier to virus spread. Alternatively, or additionally, the national parks which now contain relict assemblages of indigenous plants may be too physically remote or climatically separated from the crops that are examined for viruses. Whatever the reason, it is known that vectors are available; the impoverished insect fauna indigenous to New Zealand has been swamped by a range of introduced insects which are capable of spreading viruses.

Viruses pathogenic for vertebrates are also discontinuously distributed throughout the world. It is remarkable that India's indigenous monkey population is experimentally infectible with yellow fever virus yet remains free of infection. Similarly, whereas the introduction of exotic draft/meat animals into Africa resulted in epidemic spread of viruses that were circulating among indigenous vertebrate populations (Table 4.1), no comparable consequences for introduced agricultural animals have yet been observed in Australia.

Information about vectors is sometimes prejudiced because undue importance has been assigned to massive infestation when trivial numbers of the same species escape notice yet determine to a large extent the amount of virus spread. Taking the available evidence at face value, it seems that except when trade is involved, climate determines vector diversity, while weather determines activity and numbers which are also influenced by biotic and edaphic factors (Figure 4.1).

Data concerning the geographic/climatic range of aphids and leaf-hoppers have been collated by Eastop (in Harris and Maramorosch, 1977) and Müller (in Maramorosch and Harris, 1979), while information relating to mosquitoes, midges and other vectors of vertebrate importance has been assessed by Gillett (1971), Harwood and James (1979) and Sellers (1980). In general, the amount of virus spread is determined by the timing of vector activity (particularly migration) in relation to the availability of infectible subjects, with the temperature influencing flight and rate of moulting at the time being more important than absolute minima in northern winters or maxima during hot dry summers near the equator. However, although nematode vectors of plant viruses are not known to infest soils in the North American

Table 4.1 Some viruses naturally infecting wild animals but also liable to cause disease in introduced species

Country	Virus/disease		Reservoir invertebrate	Reservoir vertebrate	Introduced host
Africa	African swine fever	Iridoviridae	Ticks	Warthog	Domestic swine
Africa	African horse sickness	Reoviridae	Gnats: midges	Zebra, elephant	Horse
Africa	Blue tongue	Reoviridae	Gnats: midges	Antelope, wildebeest, (cattle) buffalo	Sheep
Africa	Rinderpest complex	Paramyxoviridae	None	Buffalo, numerous small ruminants	Cattle, camel, asiatic pigs
Africa	Malignant catarrh	Herpesviridae	None	Wildebeest	Cattle
Africa	Foot and mouth	Picornaviridae	None	Buffalo	Cattle
Australia	Myxoma(tosis)	Poxviridae	None	Sylvilagus brassiliensis (tropical forest rabbit)	Oryctolagus cuniculus (European rabbit)
S. America	Argentinian haemorrhagic fever	Arenaviridae	None	Calomys	Man
USA	Eastern equine encephalitis	Togaviridae	Mosquitoes	Wild birds	Pheasant

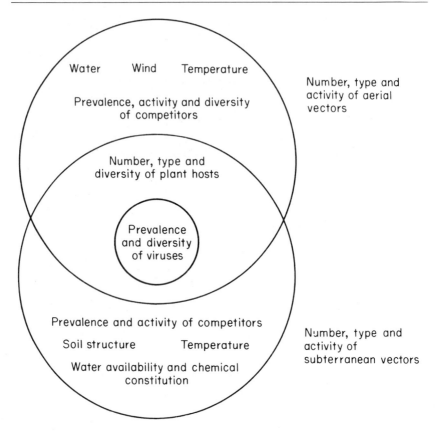

Figure 4.1 Interactions involving plant pathogenic viruses, hosts, biotic, edaphic and climatic factors.

tundra (Norton, 1978) and longidorid nematodes seem to have a pan-tropical distribution, which is reflected in the temperature optima for reproduction of the differing species, the occurrence of subterranean vectors is largely determined by properties of the soil, not temperature. In most soils, a labyrinth of spaces exists between mineral particles which are cemented by organic matter into crumbs a few millimetres or more across. Soil pores have sizes governed by the sizes of the adjoining mineral particles, and adjacent pores are connected by narrow necks whose size determines the passage of objects. Nematodes of vector genera move between particles or crumbs, rarely displacing them, and fungal zoospores almost certainly take similar paths of least resistance. The length, but particularly the girth, of nematodes in relation to the pore neck size limits movement and in the UK differences in the distribution of vector genera/species have been correlated with soil structure (Cooper, 1971). Subterranean vectors of plant viruses are also vulnerable to changes in the quality of soil water in which they swim.

Metal ions, such as zinc, kill zoospores of *Spogospora subterranea* (Cooper *et al.*, 1976), and copper kills trichodorids (Cooper, 1971). Consequently, factors such as soil pH and the prevalence of components with sequestering or buffering abilities that influence the solubility and availability of these minerals, determine whether or not a soil is liable to be infested with the subterranean vectors or, perhaps more accurately, the severity of associated disease.

4.3.2 RAINFALL

The availability of water modifies numbers and diversity of potential hosts and hence influences indirectly the occurrence of viruses and their vectors. Additionally, rainfall can have a direct effect on the amount of virus spread. Ticks seem to be indifferent but, because mosquitoes (and most other vectors of viruses pathogenic for vertebrates) lay eggs in or near water that may be more or less transient, wet seasons generally enhance their abundance and may facilitate temporary extension of their geographic range. However, when rainfall is excessive locally, giving flash-flooding or less dramatically increases river flow above a critical value which varies with the species involved, populations of vectors are minimized. Hail and rain subject insects and mites to direct physical damage, washing vectors out of the air and indeed beating them into the soil. Furthermore, aphids seem unable to walk on either wet soil or wet plants, and rainfall has been associated with considerable seasonal decline in mealybug populations (Entwistle, 1972).

Soil-inhabiting vectors require water for their movement which consequently depends on the amount and timing of rainfall (or irrigation). Excessive moisture, which is a negative factor in the survival of aerial vectors, probably does not affect fungi and nematodes directly (some of the fungi survive as resting spores and the nematodes as eggs). Nevertheless, soil saturated with water has a diminished surface/volume ratio with air and, when the oxygen concentration is depleted by aerobic microorganisms, decomposition of organic matter is accomplished by anaerobes which produce hydrogen sulphide, nitrogen, methane, etc., and in these circumstances numbers of trichodorid nematodes are known to diminish.

In soil that is not saturated, films of water surround the soil particles encompassing the large spore spaces and may fill the smaller pores. This situation favours nematode movement. However, further drying lessens the thickness of the water film, concentrates the soil solution and empties the smaller pores. Experimentally, it has been shown that virus transmission by trichodorids diminished with the soil moisture content of sandy soil (Figure 4.2), and in Europe the incidence of tobraviruses in field-grown potato tubers correlates with rainfall at

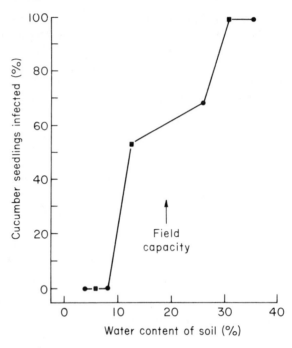

Figure 4.2 Effect of soil moisture content on transmission of tobacco rattle virus to cucumber roots by *Trichodorus* nematodes. ●, Expt 1; ■, Expt 2.

times of the year when tubers are small and infectible directly as a consequence of trichodorid feeding (Cooper and Harrison, 1973).

4.3.3 TEMPERATURE

Rainfall is inextricably associated with temperature change, and heat is critically important for arthropods and poikilothermic vertebrates, such as fish and reptiles, which lack an ability to control their body temperature within narrow limits. Virus replication is usually optimal within only a narrow temperature range, and weather has a more or less direct effect on the extrinsic incubation period in vectors. Similarly, temperature affects the activity of and hence virus transmission by passive carriers including nematodes such as trichodorids which transmit optimally at 15°C and less effectively at 24°C than 20°C (Cooper and Harrison, 1973).

Additionally, the effect of heat on virus replication in plants and animals determines the amounts available for vectors to acquire. There are many places in the world where leaf temperatures approach or exceed 37°C for a few hours every sunny day and from these plants it is possible to select thermotolerant and even thermophilic types from virus populations occurring naturally. In a series of papers published together in *Annales de Phytopathologie* (Volume 11, pp. 265–475) Quiot and his numerous colleagues reported accumulating 953 cucumovirus isolates from 39 different wild species and identified two main groups differing in thermosensitivity and other properties. Where viruses in both categories occurred together in the same plant species, their relative abundance was observed to fluctuate, with thermolabile isolates being more prevalent in France during spring and thermostable isolates being more prevalent in summer. Reflecting ambient temperature, most field isolates of cucumoviruses from the UK are of the thermolabile type, whereas isolates made in Hungary during summer are thermostable. Furthermore, temperature affects not only viruses but also the hyperparasitic satellites which they support and which sometimes potentiate their pathological impact (White *et al.*, 1995). Although the proportions of thermostable and thermolabile isolates of a virus see-saw when conditions vary, both seem to remain either as major or minor components of the virus population. However, Quiot *et al.* identified a complex of factors which interact with temperature to complicate interpretation. Similar phenomena were revealed by complementary field and laboratory experience with myxoma virus in European rabbits (Fenner *et al.*, 1974).

Even in homoeothermic animals, frosty weather lessens the skin temperature and imposes a selection pressure on viruses which are acquired and inoculated by vectors. On the one hand, rabbit fleas leave cool animals (particularly those that die), thereby increasing the prospect of transmission of isolates of myxoma virus which are not rapidly lethal. On the other hand, temperature influences the efficiency of the immune response and moderates virus replication; low temperatures tend to facilitate the acquisition of psychrophilic types of myxoma virus by vectors that feed transiently (e.g. mosquitoes). Interestingly, attenuated (less-virulent) myxoma virus isolates are less thermoresistant and low-temperature-sensitive in their replication requirements than are virulent isolates. Thus, although the viruses are equally liable to cause death, myxoma virus isolates selected for tolerance of low temperature and which take the longest time to kill rabbits, tend to become more prevalent in and after European winters.

4.3.4 NATURAL DIVERSITY AND THE BIOLOGICAL COMPONENTS

When options for dispersal are limited, the virus populations themselves tend to become adapted to the hosts and vectors available. However, given the opportunity, many viruses demonstrate their considerable variability, which allows selection for enhanced transmissibility on the one hand and the quest for new hosts on the other. Viruses adapt to change in their environment by exhibiting both polymorphism and heterozygosity (Chapter 1). However, polymorphism has compensating disadvantages; two or more viruses need to be present together and the survival of one virus may depend on another which aids its transmission by vectors. Thus, potato aucuba mosaic virus is uncommon in British potato cultivars (which are experimentally infectible) because aphids acquire the ability to transmit it only when plants infected with a potyvirus occur nearby. With a few exceptions potato cultivars grown commercially in the UK are immune to, or rapidly killed by, potato potyviruses and, in consequence, these viruses tend to be infrequent in crops. Virus transmission facilitated by transcapsidation or other forms of dependency has so far been recognized only in relation to plant pathogenic viruses having aphid or nematode vectors (Gibbs and Harrison, 1976) but it is predictable that similar phenomena operate with different sorts of viruses and vectors and may determine geographical range in some instances.

4.4 THE CONCEPT OF HOST RANGE

Virologists have revised initial ideas that a virus has only a limited host range but are unsure about the real importance of unreal experience: laboratory estimates of host range are usually based upon transmission systems that are unimportant in nature. Recognizing that wild potential hosts are usually intrinsically variable, sexually reproducing, outbreeders containing a multiplicity of virus genotypes, the concept of natural host range for any virus may be irrelevant. However, in nature, viruses undoubtedly tend to become specialized to local conditions and when these conditions remain stable, as when hosts are perennial, the associated viruses seem, when tested in the laboratory, to have narrow host ranges. Some viruses seem more mutable and catholic in their host range than others, but they also retain a genetic initiative. Different parts of long-lived plants, such as dormant buds and isolated cells in abscission zones that lie buried under tree bark for decades, contain virus isolates differing from one another in properties including temperature optima for growth and vector relations (Cooper and Edwards, 1981). Arguably, these isolates which can be distinguished in laboratory tests have been selected to optimize use of conditions that pre-exist in nature and remain in anticipation of a return to those conditions.

When natural hosts are ephemeral there is a greater tendency towards polyphagism by viruses. Alfalfa mosaic ilarvirus and tobacco rattle tobravirus are among the most opportunistic; each experimentally infects hundreds of species in dozens of plant families (for example, 150 species in 22 families are infectible by alfalfa mosaic virus under natural conditions). Few studies have been made in natural communities but, in one that was done in the USA (Bald, 1960) it was found that two solanaceous species contained distinguishable to-bamoviruses which were nevertheless able to cross the species boundary as a result of experimental challenge. The use of targeted mutation and transcripts from DNA copies of viral genomes ('knock-out' mutants) is tending to suggest that the host genotype is a crucial variable and this knowledge might have important implications when optimizing the development and durability of a new range of transgenic resistance (even immunity) genes. However, reliable data obtained with a mutant in one plant species are not necessarily reproducible in another. An implication may be that a large number of lock and key interactions are possible between viruses and their hosts. On this basis, experience suggests that some species (e.g. *Stellaria media; Chenopodium quinoa*) may have a greater number of 'locks' than others (i.e. they are charac-terized by their infectibility by diverse viruses in nature); on the basis of experimental infectibility, *Nicotiana benthamiana* might be in the same category. The picture is not clear at present because viruses undoubtedly differ in unappreciated and poorly understood ways that impact upon such interpretations. Thus, to take one unexpected example, coat protein seems important for the systemic invasion of comoviruses and bromoviruses but not hordeiviruses or geminiviruses. There is a paucity of critical data on which to base generalization but if true, this might mean that viruses are to some extent restricted and natural hosts in one taxon are insignificant sources of viruses for plants in others – perhaps even closely related individuals within a species. Plasmodiophorid fungi which have persistent vector relationships with viruses pathogenic for spermatophytes probably support the replication of the viruses they carry and undoubtedly pose a threat to higher plants. Other taxa, such as prokaryotes or algae, though largely un-tested, cannot yet be considered sources of viruses for higher plants. Cycads and gymnosperms are naturally infectible with viruses that also infect broad-leaved angiosperms (Cooper, 1979, 1993) but it is note-worthy that there seems to be a sharp distinction between dicotyl-edonous and monocotyledonous herbs which have been studied in greater detail. With a few exceptions, viruses infecting monocotyledons do not systematically invade (and indeed seem not to infect) dicotyledons either naturally or experimentally.

Among viruses that infect vertebrates, the humoral components of 'immunity' apply selection pressures different from and additional to those that are applied against viruses that infect plants (although it is likely that antibodies will be expressed in plants to 'protect' them from viruses; *viz.* Taviadoraki *et al.*, 1993). Irrespective of the longevity of vertebrates, the more or less durable components of humoral 'immunity' usually determine the extent to which an animal is invaded.

The natural host range of viruses that infect vertebrates is imperfectly known, but host restriction to one or more closely related species seems uncommon, although not unknown; rubella (Togaviridae), varicella (Herpesviridae) and measles (Paramyxoviridae) have humans as the main natural host. Indeed, poxviruses seem to have very restricted natural and experimental host ranges. Avian isolates of poxviruses which have been obtained from tree creepers, crows, finches, mockingbirds, pheasants, woodpeckers, starlings, grouse and thrushes only very infrequently infected domestic poultry and poxviruses from poultry did not infect twelve species of wild birds (Davies *et al.*, 1971). Newcastle disease virus (NDV, Paramyxoviridae) is at another extreme. A variety of gallinaceous species as well as sparrows, pheasants, pigeons, crows, parrots, ostriches, vultures, shags and cormorants are naturally infected by NDV (Lancaster, 1966; Lancaster and Alexander, 1975). Furthermore, land mammals have been implicated as additional hosts. Indeed, since rats are infectible, as are cats *per os*, it is possible that the cat–rodent–cat infection cycle plays some part in NDV maintenance at poultry farms during the periods of 'changeover' which the disease periodically forces poultrymen to adopt to minimize commercial losses.

Human enteroviruses, reoviruses, togaviruses, rhinoviruses, paramyxoviruses and adenoviruses have antigenic analogues that naturally infect wild and domesticated animals, but, recognizing the extreme genotypic and phenotypic variation that is possible with viruses, it may be unwise to infer for example that the enteric picornavirus designated Coxsackie B_5 which can be isolated from humans was the progenitor of the agent causing swine vesicular disease, even though these viruses have numerous properties in common. In the same way it is unreasonable to infer, as Johnson *et al.* (1969) did, that rhabdoviruses such as cause vesicular stomatitis in cattle are transvestite plant pathogens which also infect invertebrates. Invertebrates are well placed to have been (and be) one of the melting pots for virus evolution; close similarities between plant-infecting tospoviruses and vertebrate bunyaviruses is consistent with this possibility. However, the ultimate origins of viruses are unknown and almost certainly very diverse. Indeed, there are few viruses for which even the immediate sources of inoculum are known or suspected (Table 4.2).

4.5 FACTORS FACILITATING ACQUISITION OF NOVEL HOSTS

4.5.1 YELLOW FEVER

Awareness of the possibility that viruses circulate within wild populations and may spill over to involve domesticated animals, humans or their crops was largely created (*c.* 1930) as a consequence of detailed

Table 4.2 Some known sources of virus causing animal disease

	Virus or disease	Source	Route of inoculation
Poxviridae	Cowpox	Cattle	Skin abrasion
Paramyxoviridae	Newcastle disease	Birds	Conjunctival contact
Orthomyxoviridae*	Influenza A	Horse, swine, birds	Respiratory
Picornaviridae	Foot and mouth	Elephant, buffalo, deer	Respiratory
Rhabdoviridae	Rabies	Canines, felines, cattle, bats	Vertebrate biting
Arenaviridae	Lymphocytic choriomeningitis	Rodents	Respiratory

* see p. 114 for qualification.

research into yellow fever virus (Togaviridae). Complementary studies in Africa and South America emphasized the role of wild animals, in this instance primates (with mosquitoes), as reservoirs of infection for humans in and near tropical rain forests (Strode, 1951; Schlesinger, 1980).

Traditionally, humans have intruded into virgin forest for hunting/woodcutting, but now oil prospecting and the exploiting of mineral resources justify such incursion. The main impact of yellow fever disease was felt in South and Central America where the virus showed itself to be lethal for monkeys and marmosets between which it is spread by mosquitoes, notably *Haemagogus spegazzinii* (Figure 4.3) and to a lesser extent *Aedes leucocelanus*. These insects, like the primates they pester, are for the most part arboreal. Possibly, there are additional cycles of infection involving marsupials, rodents and other vectors. However, when the forest trees are cut down, woodcutters are exposed to clouds of mosquitoes from the canopy, are bitten and may be infected with yellow fever virus.

About 5 days following infection in humans, and somewhat less in monkeys, yellow fever virus is present in the peripheral blood and remains at a high concentration for 2–3 days, whether the disease is overt or silent. Thus, virus is accessible to other feeding mosquitoes but only briefly and the ambulant human patient typically serves as a link connecting the forest with his or her urban environment where mosquitoes of other species and genera, particularly *Aedes aegypti*, abound in all domestic rubbish containing water. In Africa, where the virus was not definitely known to occur until 1927, the situation is different (Figure 4.4). The wild primates, not humans, act as porters of inoculum between the sylvan and urban environments.

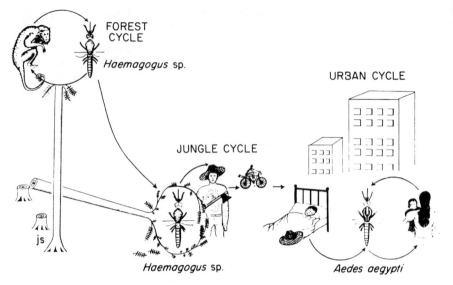

Figure 4.3 The role of wild animals (primates plus possibly marsupials and rodents) as reservoirs of yellow fever virus in South America. Forest workers infected with virus carried by *Haemagogus* mosquitoes may bring infection to villages where the virus is disseminated by other vectors (notably *Aedes aegypti*).

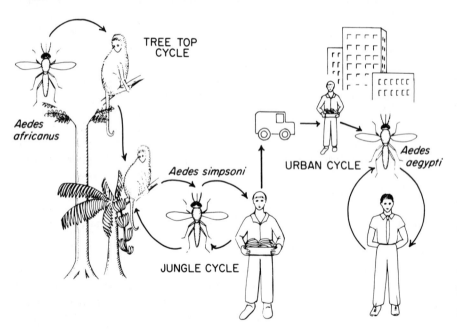

Figure 4.4 The role of wild animals (chiefly primates) as reservoirs of yellow fever virus in Africa. In the forest, monkeys maintain the virus and *Aedes africanus* is an important vector. This mosquito may infect man directly but the virus is usually introducd to urban man in a different way. Monkeys feeding in periurban plantations are exposed to *Aedes simpsoni* which thereafter facilitate dispersal within the human population. *Aedes aegypti* is in Africa and is liable to be an important virus vector both in the forest and in villages.

However, mosquitoes are also involved as vectors and monkeys are also prime maintenance hosts of the virus in this continent. Thus, it is well established that, in Africa, Colobus monkeys in particular host the virus in the tree tops where *A. africanus* act as a prime vector. However, both co-habit with more venturesome primates, e.g. red-tailed guenon (*Cercopithecus ascanius schmidti*), liable to invade human settlements to feed on banana or other crops. Here, the monkeys are exposed to *A. simpsoni*, which breeds in leaf axils of plants, and is therefore well placed to spread virus to humans who will return to their urban surrounds where *A. aegypti* is again the most frequent vector. Plausibly with the expansion of timber industries in Africa, *A. africanus* will be more frequently able to inoculate humans directly. However, changed behaviour of the virus with possibly a new vector and reservoir other than monkeys has been observed in Colombia where farmers who had no contact with the forest have been infected with yellow fever.

Because there is only transient human carriage of virus in blood, maintenance depends upon the extrinsic mosquito vertical components. Even though mosquitoes remain infected for life (months), the transmission cycle of yellow fever virus in humans is usually broken unless fresh human infectibles are available. These factors combine with the human mortality characteristic of the virus (5–85% among frank cases) to give yellow fever dramatic epidemic potential, but rarely, if ever, endemic status, in humans. In non-human primates, the virus is undoubtedly equally liable to flare up and burn itself out but the faster breeding rate and rapid mobility of monkey populations facilitates virus dispersal on the one hand and on the other leaves immune survivors of infection capable of restoring the numbers of infectibles, thereby allowing maintenance. It is likely that only the three or four greatest forests, notably those of Amazonia, provide sufficient continuity in space to support an endless reach for non-immune hosts and in time to allow regeneration of the essential individuals without previous exposure. However, the behaviour of the virus in Senegal and Gambia has led WHO to conclude that yellow fever virus can survive during prolonged dry seasons in eggs of *Aedes aegypti*. Fortunately, transmission of yellow fever virus by aerosols, which has been demonstrated experimentally, is not known to occur in nature. Consequently, in temperate regions of the world where winter temperatures are inimical to vector (*A. aegypti*) survival, yellow fever is an exclusively summer event which is contained and very uncommon because of vaccination and rigid quarantine. During the nineteenth century, however, shore-based labour, infected ships' crews and/or mosquito flight caused numerous introductions into Europe and the main ports of the USA.

Thus, even when viruses have broad natural host ranges, they may remain localized, especially when their vectors have exacting maintenance requirement. At one extreme, the bunyavirus causing California encephalitis characteristically infects children who climb trees. The mosquito vector (*Aedes triseriatus*) typically breeds in water that accumulates at the forks of trees, maintenance of the virus in these circumscribed sylvan situations being helped because of transovarial transmission in the vectors. Picnicking and recreation apart, opportunities for human contact with this virus are rare but massive incursion into hidden foci of infection sometimes results in complicated interactions with indigenous fauna which results in epidemics.

4.5.2 KYASANUR FOREST DISEASE

Kyasanur Forest disease virus (KFDV; Togaviridae), which is responsible for haemorrhagic disease in primates, illustrates some of the factors contributing to epidemic spread. Fortunately, even though the virus is endemic, it has remained localized within marshy scrub communities from Europe to Asia, and only occasionally reveals itself by causing encephalitis in foresters and woodcutters (Work, 1958). Although serological surveys have implicated numerous small mammals as natural KFDV reservoirs, an abnormally high incidence of mortality in monkeys in Kyasanur Forest, India, justified detailed investigation. Following a doubling of the human population near Kyasanur Forest, KFD appeared as an apparently wholly new disease. It was rapidly established that the greatly increased human population has consequential effects on the sylvan environment due to increased cattle and food crop production coupled with the additional need for structural timber/firewood (Boshell, 1969). Human incursion into the forest was also necessary to obtain green manure for betelnut (*Areca catehu* L.) cash crops and the fertilizer/mulch was obtained by pruning forest vegetation. Land cleared in this way was rapidly invaded by *Lantana* spp., plants which had been introduced into Ceylon from tropical America for ornamental planting but escaped to form dense, sometimes impenetrable, thickets harbouring dispossessed refugees from the original forest fauna. Research revealed that the human and indeed the monkey infections with the agent of KFD were mere indications of an inapparent epizootic (an epidemic in wild animals) of rodents and jungle fowl (Figure 4.5).

The lantana provided cover and food for numerous birds such as *Gallus sonneratii* and small mammals, notably mice (*Mus booduga*), shrews (*Suncus murirus*) plus, probably to a lesser extent, *Rattus* spp. Given the changed conditions, pests prospered and polyphagous blood-sucking Ixodid ticks, particularly *Haemophysalis* spp. rapidly exploited the new habitat with its augmented forest floor fauna. Humans introduced additional factors into the already complex situation;

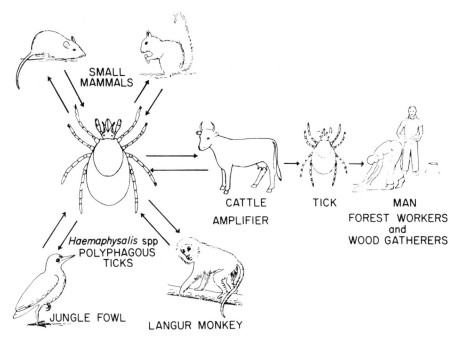

Figure 4.5 The roles of wild and domestic/commensal animals as reservoirs of Kyasanur Forest disease virus. Human infection is consequent upon three interacting factors. (i) Exploitation of forest by humans and their animals; (ii) invasion of cleared forest by *Lantana* sp. providing cover and food for additional small mammals and birds; (iii) greatly increased numbers of polyphagous blood-sucking tick vectors (*Haemaphysalis* spp.) exploiting the augmented fauna. Cattle and goats amplify the vector population, acting also as 'porters' facilitating vector contact with human hosts.

foraging cattle and goats provided sources of blood and further amplified the massive tick infestation. These animals acted as porters of virus-carrying vectors which were thereby given the opportunity to feed on human blood. It is probable that in this locality the ground fauna acted as the main reservoir of both virus and vector; arboreal monkeys may have had infection brought to them by langurs (*Presbytis entellus*) and other primates, e.g. *Macaca radiata*, which spend part of their lives on the ground. Primates are 'dead-end hosts' incidental to the viraemic core of the KFD virus maintenance cycles and the same is probably true for the hantaviruses which periodically reveal themselves in human populations that impinge upon infected rodents (Nerurkar *et al.*, 1994). However, numerous viruses that have been similarly encountered have been capable of maintenance in exotic hosts. A suite of mosquito-borne togaviruses causing dengue fever have monkey maintenance hosts in Malaysia but have shown a marked independence

from the presumed primate reservoir: Man–mosquito–man transmission cycles are normal and these viruses are an omnipresent threat
to human health wherever *A. aegypti* occur.

Plausibly, the natural inhabitants of regions where viruses are
localized have been selected for tolerance but nevertheless become
infected and consequently exhibit immunity which provides additional
containment. However, outbreeding populations are rarely uniform in
their patterns of immune responsiveness. Thus, among cattle and wild
ungulates the immunity to foot-and-mouth apthovirus (Picornaviridae)
that follows infection is thought to wane rather rapidly, at least in a
few animals because re-infection with homologous types of the same
virus can occur and produce enough virus in the associated lesions to
infect other animals. The intrinsic variability of populations facilitates
virus maintenance but may force changes in the antigenic properties of
virus particles. This may be an important evolutionary mechanism in
some viruses of long-lived vertebrates within which infection is more
or less restricted to a mucous surface such as the respiratory epithelium
or the gut.

Influenza A (Orthomyxoviridae) has been a notable model (Stuart
Harris and Schild, 1976), and it is significant that the IgA against any
one antigenic form of the virus persists in the respiratory tract at
protecting levels only for about 5 years, not for life. Recurrent infection
with obvious signs of disease in humans is usually attributable to a
series of antigenically distinct serotypes of the virus, a phenomenon
termed antigenic drift. Although influenza A viruses have only been
available (cultured) for detailed study since 1933, four major and several
minor antigenic changes have been recognized in the natural virus
population circulating among humans. Each major change nullified
pre-existing immunity within the human population and has been
associated with the appearance of worldwide (Figure 4.6) disease
epidemics (pandemics).

The question of fundamental importance which has so far not been
answered satisfactorily is that of the origin of the new subtypes. Two
main explanations have been suggested.

1. Human pandemic influenza A viruses arise by mutation of pre-
   existing types. Little biochemical evidence favours this possibility
   and the hypothesis is not widely supported at least in part because,
   despite immense opportunity and untold millions of cycles of virus
   multiplications, changes in antigenic composition have been recorded
   only once during influenza epidemics (Yewdell and Gerhard, 1981).
2. The more popular view is that pandemic influenza A viruses arise
   directly or indirectly through recombination. Influenza A contains
   six genetic linkage groups and experimental data show that these

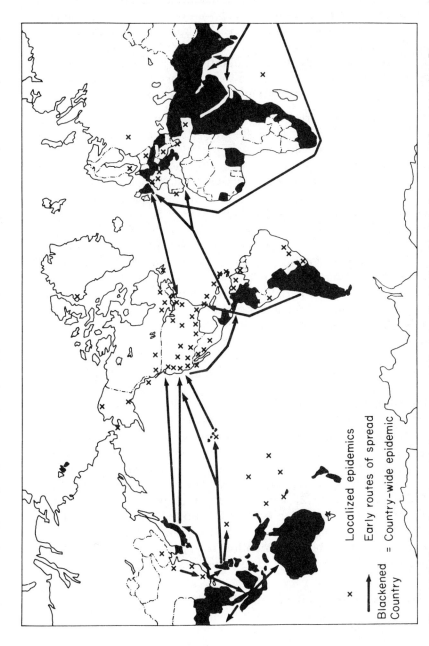

Figure 4.6 The early routes of influenza spread that resulted in localized and country-wide epidemics. Reproduced from Rivers and Horsfall (1959) with permission from US Department of Health, Education and Welfare, Public Health Service.

are amenable to redistribution between particles having different properties. A diverse array of antigenically distinct influenza A viruses exists in non-human primates and other vertebrates, including birds. Significantly, genetic reactivation and resortment occurs readily in the respiratory tracts of chicken and swine. Genome breakage and repair (recombination) has not been observed but pseudorecombinants (re-assortments) occur with frequencies approaching 95% between certain influenza viruses.

Swine in several countries of the world have been naturally infected following contact with human influenza A virus and it is noteworthy that some pigs carry and shed virus for periods of months. Interestingly, in humans, each successive antigenic shift tends to result in the exclusion of pre-existing viruses (although this does not necessarily happen in other animals) and is brought about because recurrent exposure to and infection by viruses having antigens in common tends to enhance the specific antibody response to the first virus in the series. This phenomenon complicates the interpretation of retrospective serological surveys involving paramyxoviruses and arboviruses (notably togaviruses), making it impossible to identify the particular serotype to which vertebrates were earlier exposed. However, because sequential infections additionally broaden the antibody spectrum of their hosts, there is a tendency for enhanced protection to a suite of viruses having even a few common antigenic determinants. In most instances, this second property of humoral responsiveness merely lessens the severity or duration of the associated disease; it has only a slight effect on the amount of available inoculum.

Though unproven, it is a virtual certainty that human influenzas are recent diseases of humans, the viruses being derived from wild forms of life. Since many species of birds are chronic shedders of influenza A viruses throughout their lives (virus is present in droppings liable to contaminate food, water and pastures) the avifauna is usually considered to be the most probable natural reservoir (Laver and Webster, 1979; Gorman et al., 1992). However, although viruses associated with pig flu in Europe have avian affinities (Schultz et al., 1991), studies of porcine flu evolution in Southern China revealed no evidence of gene flow from avian flu but did identify interspecies gene flow between viruses of humans or pigs – and re-assortants (Shu et al., 1994).

4.6 THE BROADCASTING OF VIRUSES

4.6.1 IN THE AIR

Being infectible by numerous viruses, wild birds are known or strongly suspected to facilitate transcontinental virus dispersal: they also provide, at least occasionally, means whereby vectors and virus-infected seeds can be dispersed over long distances (van der Pijl, 1972). Transport of eastern equine encephalitis (Togaviridae) from North America to the tropics is undoubtedly attributable to birds which also contribute to the movement of the virus within the USA (Hayes and Wallis, 1977; McLintock, 1978). In nature, a few wild vertebrates other than birds seem, at least locally, to be important agents of virus dispersal. Bats, which are notable in this context, are second only to rodents in their diversity: they are among the most widely distributed, abundant and gregarious of mammals. Millions of bats co-habit in caves, hollow trees, etc. with similarly large populations of invertebrate parasites. Bats, which hibernate when aerial vectors and alternative hosts are few or absent, live for up to 20 years and plausibly act as maintenance hosts for some of the mosquito-borne togaviruses (Smith, 1989). Furthermore, bats are reservoirs for a suite of closely related rhabdoviruses including the virus causing rabies and since bats have been known to bite people who picked them up, it is likely that wild carnivores are from time to time infected by rabies virus in similar circumstances. Usually, fruit-eating and insectivorous bats avoid contacting humans, but this contrasts with the behaviour of vampires (*Desmodus* spp.) which routinely feed on mammalian blood. Bites from vampire bats occasionally incite rabies in humans, although terrestrial wildlife, marine mammals breeding ashore and domesticated stock are the more usual victims of their stealthy attacks.

4.6.2 ON THE WIND

People who enter caves where rabid bats roost have enhanced exposure to inoculum in the air, and, in one or two instances rabies seems to have occurred as a result of virus inhalation. Most viruses that spread by the respiratory route rapidly cease to be infective when dried in air, but the effects of light and ionizing radiation may also be important (Chessin, 1972; Killick, 1990). However, in moist air, the viruses may be carried great distances. Dissemination of foot-and-mouth disease virus (FMDV) over 50–100 miles has been plausibly attributed to wind, although the activities of birds cannot be excluded. In Denmark, FMDV was wind-borne for up to 18 miles over the sea and exceptionally over twice this distance, darkness and high relative humidity favouring prospects for spread (Donaldson and Ferris, 1975). Subsequently, meteorological data (principally air trajectories) strongly implicated a pig unit in France as the source of inoculum for FMDV in the Isle of Wight, the Channel Islands (Jersey) and in Brittany (Figure 4.7; Donaldson et al., 1982). The salt and protein content of the material in which viruses are conveyed and the extent of atmospheric pollution

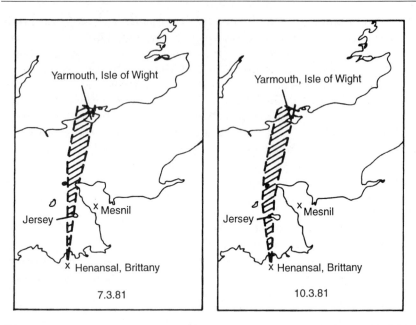

Figure 4.7 Air trajectories from Brittany to the UK on March 7 and 10, 1981. Reproduced from Donaldson *et al*. The Veterinary Record (1982) **110**, with permission.

probably influence the survival of infectivity. Thus, whereas heavy rainfall aids virus deposition on the one hand, it tends to clear the atmosphere of materials harmful to viruses on the other, thereby prolonging survival prospects. In contiguous poultry units having forced ventilation, there is a proven risk of 'artificial' wind-assisted dispersal of Newcastle disease virus (Hugh-Jones *et al*., 1973).

The wind is a well-documented aid to the dispersal of invertebrates. Aphids have been blown from mainland Eurasia to Spitzbergen (*c.* 400 km), and infestation of the northern islands of Japan from the Peoples Republic of China seems to be commonplace. When journeys have been made overland, it is rarely possible to be sure that only one generation of insects was involved; most insects fly (or are wind-assisted) in a series of short hops. Indeed, insects carried on surface winds are often deposited near their source in regions of local turbulence caused by topographical features or, more important in a crop context, at boundaries between fields containing plants of differing size (Lewis, 1966). However, when wind-assisted, distances covered are commonly about 5 km/day and, when the insects are carrying viruses, this dissemination may provide important sources of inoculum for vertebrates

(Sellers, 1980) or plants. Until recently, virologists assumed that vectors would not spread non-persistent viruses except locally. However, records of myxoma virus introduction into island populations of rabbits suggest that 100–200-km journeys are possible, even though this virus is a mechanical contaminant of flying pins. Furthermore, the sudden appearance of maize dwarf mosaic potyvirus in the mid-western states of the USA, after a journey of 800 km in jet-stream winds at about 80 km/h, has called into question the relevance to natural situations of laboratory determinations of virus retention (Berger and Zeyen, 1981). Maize dwarf mosaic virus can be carried by aphids that retain their ability to transmit for a maximum of 6 hours at 20°C, but temperatures of 7°C diminish rates of infectivity loss and extend retention times to 70 hours. At the altitudes in which insects travel long distances in air, local temperatures are probably low enough to extend periods of virus retention over those usually quoted.

These natural means whereby viruses and their vectors may be carried in air are augmented to an important extent by aircraft movement. There is strong circumstantial evidence that aphids of North American origin were transported to south-east Asia by air in time of war, and there is widespread concern about the risks of introducing viruses in similar circumstances. The speed of transportation allows waste food originating in one country to be sufficiently turgid when discarded in a second country (perhaps in another continent) to be fed upon by aphids that may acquire abilities to transmit new viruses as a consequence. Of course, there is now a routine trade by air in 'perishable' produce which provides out of season choice. From time to time, such material will be available on market stalls for virus acquisition by vectors. Analogous importations of inadequately cooked meats (such as salami) are known to be potent sources of contagion for swill-fed pigs and waste food from ships has long been recognized as a source of foot-and-mouth disease virus responsible, for example, for introducing the virus into California on at least two occasions. For these reasons, countries that are free from viruses such as cause foot-and-mouth disease or swine vesicular disease legislate and educate, and publicize the specific risks associated with refuse from ships or aircraft. Governments also advise about the hazards attributable to the inadvertent introduction of exotic pathogens, but the pressures of trade, and hence science, are such that the hazard may be increasing rather than diminishing.

4.6.3 VIRUSES BROADCAST BECAUSE OF COMMERCE OR SCIENCE

Infected plants usually carry viruses for life, and because the likelihood has long been recognized that commercially propagated plants contain viruses, expensive steps are commonly taken to minimize the risks attendant on trade, whether national or international. Most countries seek to control the amount of imported plants or soil while monitoring the entry of packaging material and other plant products (Hewitt and Chiarappa, 1977). Unfortunately, not all countries that have specialists with the time, technology and expertise required to identify viruses publish the results of tests made. However, to illustrate the magnitude of the problem it is worth noting that one institution in the USA reported 62% incidence of virus-like agents in 1277 vegetatively pro- pagated plant introductions received between 1957 and 1967. Over a similar period when 551 plant introductions of vegetative propagules (mostly tubers of the genus *Solanum*) were tested, 73% (of 445) cul- tivated and 39% (of 106) wild specimens contained one or more viruses. Tests in Europe are consistent; Bode (1977) tested 930 tubers of *Solanum* spp. and found only 5% to be free of viruses: multiple infection was common and fewer than 30% contained only one virus. Much of this material was introduced into Europe or the USA for scientific research. When assembling wild specimens for use in crop breeding or for taxonomy, collectors have some knowledge of the potential pathogens and tend to select plants free from foliar blemishes. Consequently, they are liable to facilitate the movement (usually internationally) of viruses that are not apparent in the collected individual. However, inapparency need not characterize the interactions of these viruses in other sorts of plants. In many instances horticulturalists have different objectives from those of scientific collectors. For example, plants with yellow variegated foliage are highly prized by gardeners and wild hosts of viruses that cause bright yellowness in leaves are more likely to be collected than those containing viruses associated with mild or transient effects. This might explain why more than half of the known tymo- viruses (which seem uncommon in natural plant populations) have nevertheless been recognized in botanic garden specimens rather than the more closely monitored food crops. All but a few of the commerci- ally important viruses pathogenic for potato and *Spongospora subterranea*, the vector for potato mop-top furovirus, seem to have come to countries in which *Solanum tuberosum* is a food crop, with planting stock that originated in centres of *Solanum* species diversity such as the Andean region of South America. Similarly, the transportation of phylloxera (aphid)-resistant grapevines facilitated the spread of grapevine fan leaf nepovirus and its nematode vector (*Xiphinema index*) from a common locality in Asia Minor (Martelli, 1978). Furthermore, the most important vector (*Circulifer tenellus*) of sugar beet curly top geminivirus in the USA seems to have originated in the Old World (Mediterranean and Asia Minor). Bennett (1967b) suggested that fodder beet for animals carried on ships by immigrants at the time of the

California gold rush could have transported the virus while also supporting colonies of the vector. Analogously, *Frankiniella occidentalis*, the highly efficient thrips vector of tomato spotted wilt tospovirus was recently introduced into Western Australia to refresh an epiphytotic for which indigenous thrips were inefficient vectors (Malipatil *et al.*, 1993).

Viruses pathogenic for vertebrates have also been broadcast in commerce. Thus, it is tempting to infer that yellow fever virus, along with its most important urban vector (*Aedes aegypti*), was introduced as a consequence of the slave trade, to the neotropics, where the virus decimates primate populations, from Africa, where it does not. Pilgrims, refugees and guerrillas have also been implicated occasionally as porters of viruses pathogenic for humans, and pastoral nomads moving herds of food animals overland routinely are major factors in the maintenance or spread of rinderpest in Africa. Transhumance has introduced Rift Valley fever virus into Egypt, broadcast foot-and-mouth disease in India and its modern equivalent (tourism) has probably been responsible for localized evolution of Ross River alphavirus (Lindsay *et al.*, 1993).

Scientific investigations aimed at optimizing food production in an overpopulated world have, as previously mentioned, facilitated the international movement of plant pathogens, but scientists have in a few instances also exposed themselves and others to a diverse array of virus-like agents pathogenic for vertebrates (Collins, 1983). Animal skins used in taxonomy sometimes have viable ticks liable to carry viruses attached to them. Even laboratory chemicals may be contaminated, e.g. commercial sources of the enzyme trypsin have been shown to contain porcine parvoviruses (Croghan *et al.*, 1973). Living specimens introduced into zoological gardens have on numerous occasions been the source of infections that devastated the captive population. The viruses of foot-and-mouth disease, rinderpest and Newcastle disease are notable in this context because they spread uncontrollably in air and infect through the upper respiratory tract. However, it is interesting that togaviruses, which are maintained naturally in non-human vertebrates and usually require invertebrate vectors, are nevertheless frequent causes of disease and occasionally death even for those working in well-equipped virological laboratories.

Wild-caught animals are a hazard. The risks they present are probably in proportion to exposure, i.e. scale- and time-dependent but difficult to quantify except in retrospect. It is, however, pertinent that thousands of rodents and wild primates are used each month for a range of laboratory tests. These collections of experimental animals increase dramatically the possibilities for disseminating viruses among human communities (Hull, 1968; Armstrong *et al.*, 1969). Some, such as the herpes virus B from rhesus monkeys, are typically lethal for humans, and others have been associated with a disturbingly high mortality (c. 25%) among infected laboratory and medical staff in Europe. The immediate source of the Marburg filovirus was monkeys imported from Uganda, but monkey infection may have occurred in transit when numerous primates were housed near other vertebrates. Because of isolation and awareness, the Marburg filovirus was not broadcast widely within the human community and, partly by luck coupled with drastic intervention, another primate filovirus was contained in the USA (Preston, 1994). However, it is noteworthy that humans have unintentionally (but on a massive scale) disseminated other viruses (notably simian virus 40) from primates (Chapter 1, p. 16).

4.7 EPIDEMICS, PROSPECT AND RETROSPECT

Primitive humans survived periods when they domesticated the animals on which they now depend, but the increased management of wild animals for human food that is anticipated, notably in Africa, will inevitably enhance the prospects for viruses to contact the immunologically unprotected. Large-scale interference with the terrestrial environment is probably essential if the food needs of the world's escalating populations are to be satisfied. However, the greatest care should be exercised when exploiting virgin terrain. There is widespread concern that the Pan-American highway will facilitate the mingling of viruses (indigenous to North or South America) which have hitherto been isolated by the sea, ice or the inhospitable rain forests of Colombia. This enhanced opportunity for genetic interaction allows viruses to acquire new hosts and the consequences cannot be predicted accurately, but the terrestrial environment which is presently dominated by humans is very vulnerable to epidemics.

Subsistence production of crops depending on a large number of wild plants or animals reared in mixture with superior, but nevertheless genetically heterozygous, populations has been largely superseded. With the notable exception of farming in tropical forests where crops such as cocoa are grown under high forest and managed or neglected more or less in line with the fluctuating commercial value of the produce, fields are no longer small, scattered and abandoned for periods greater than they are cropped. Now, the species diversity of the flora is diminished and the range of crop genotypes is fewer because harvests are synchronized. Two crops, wheat (virtually continuous over North America and similar climatic zones of the former Soviet Union) and rice (which covers most river valleys of Asia and Oceania) provide food for at least half the world's population while sugar beet with potatoes are almost as widespread. In these crops the diminishing genetic diversity is selected. By contrast the narrow genetic base in the yams and plantains which are stable foods in tropical regions is a result of intrinsic constraints. To optimize use of land, crops are also grown in partially overlapping sequence, and arid regions that were hitherto unusable have been brought into cultivation as a result of irrigation. Inevitably, these factors greatly enhance the prospect of exotic plants acquiring viruses from remnants of the indigenous flora, and epidemics have followed (Thresh, 1980a, 1982). There have also been less predictable sequelae. Many hitherto unrecognized mosquito-borne arboviruses that naturally infect birds are now more prevalent because water-filled irrigation ditches provide sites for vector multiplication. Furthermore, increased crop density and yield has encouraged the build-up of rodent populations carrying arenaviruses with a proven potential for uncontrollable spread in air to cause fatal haemorrhagic fever in humans. Some 100 000 km^2 of Argentina's richest agricultural land adjoining almost half of the human population of that country is known to be exposed to this contamination, where species of *Calomys* seem to be the only important known reservoir rodents. However, elsewhere in South America, another species (*Calomys callosus*) which has a marked tendency to invade urban environments has been implicated as a reservoir for another arenavirus (Machupo). The prospect is disturbing that different rodents commensal with humans, such as *Rattus rattus*, might become infected when foraging on land before returning to ships which could facilitate the dispersal worldwide of these South American arenaviruses.

4.7.1 MORBILLIVIRUSES AND PARVOVIRUSES – RINGING THE CHANGES

The spread of viruses within mobile populations of infectibles is undoubtedly helped when they are gregarious. In nature, this happens at waterholes and while feeding possibly for mutual protection, to confuse predators or to facilitate hunting, as with wolves. Human populations are routinely gregarious but periodically assemble to an even greater extent than normal. Analogously, domesticated animals are brought together for convenience, safety, sale, shearing or show, and each of these potentially facilitates virus transmission. The exceptional mingling of diverse military populations undoubtedly contributed to the 1918 pandemic of influenza A and the epidemic of paralytic poliomyelitis in the UK that followed World War II. Analogously, military campaigns since the time of Charlemagne have allowed the pandemic spread of rinderpest which has decimated livestock numbers in Europe (Fleming, 1871, 1882) and Africa from a base in Asia. Following introduction of the virus into Africa up to 90% of domestic cattle and wild buffalo were killed in and soon after 1889.

Circumstantial evidence (Fraser and Martin, 1978) is consistent with human measles virus having evolved from a suite of morbilli viruses (Gibbs *et al.*, 1979) prevalent in canines (causing distemper) and wild ungulates. Although these are antigenically similar to one another, phocid distemper (Chapter 5) is distinctive and more similar to canine distemper virus. There is no way to deduce whether human measles virus was the progenitor of these agents or whether rinderpest was the forerunner in evolutionary terms. The dolphin and porpoise morbilliviruses seem to be recently evolved variants that share many molecular features, but morbillivirus introduction into land-based communities as a result of marine movements is an ever-present possibility. The management of epidemics usually centres on control of trade in live terrestrial animals. Of these, subclinically infected pigs, sheep and goats pose a threat but the maintenance of morbilliviruses largely depends on mild genotypes of the viruses (including attenuated goat-adapted rinderpest vaccine isolates) and wildlife reservoirs – from small ruminants to hippopotamus. Canine distemper is an important factor in wildlife ecology and has been implicated in the virtual elimination of a species (black-footed ferret) in North America. Unexpected impact on wildlife is a complication in the development of vaccination programmes aiming at managing losses in domestic stock in Africa and Asia. Thus, avianized canine distemper vaccine genotypes are severely damaging in rare and valued species such as the giant panda – yet completely satisfactory for the protection of canines.

The aggregation of inbred populations of domesticated animals (Biggs, 1978) provides excellent conditions for the spread of initially uncommon viruses (Hugh-Jones, 1972), although it also facilitates protective vaccination. Parvoviruses which occur worldwide have demonstrated dramatic epidemic potential. The virions are durable and their persistence is facilitated because latency, periodic excretion and venereal transmission complement the 'normal' aerosol or faecal–oral routes of dissemination. During 1978, canine parvovirus became pandemic (even invading countries such as Australia which have strictly enforced quarantine policies for canines). Over a 3-year period, a very small number of base changes (perhaps as few as two) in a progenitor virus (such as has been characterized from racoons and cats) seems to have been responsible for the pandemic in dogs (Parrish *et al.*, 1990).

The sort of transformation that has endowed parvoviruses with a broadened host range or other viruses with the capacity for man–man rather than wildlife–man transmission tends to liberate the agents from climatic/geographic constraints imposed by their earlier hosts. The triggers for such changes remain unexplained but are likely to recur and, whenever a virus acquires a new host, the consequences are difficult to foresee. Despite man's incautious efforts, all viruses are not yet everywhere.

4.8 A NEW DIMENSION – GENETICALLY ENGINEERED VIRUSES

Transformation and transduction of bacteria lead the way into gene delivery systems and derivatives of the bacteriophage lambda were among the first used for this purpose. For about 15 years, a diverse array of laboratory-made recombinant viruses has been used to address aspects of cell regulation but also opportunities for delivery of virus-derived genes/products into plants or animals. Most resources are devoted to the development of vaccines for use in human or veterinary medicine and, in this context, retroviruses have been engineered to address gene delivery into specific blood cells (CD4+). Despite some structural limitations, retroviruses are among the most efficient ways of integrating extraneous DNA into vertebrate cells where they can facilitate targeting of chemotherapeutic agents in cancer treatment or gene therapy (gene function replacement). Inevitably this work has a long gestation period and was, until recently, restricted to laboratories. However, a few genetically engineered viruses have been authorized for experimental use in the field (for biological control of insects or vaccination of wildlife).

Because of environmental awareness and evolution in the target pathogens and pests, the range of effective pesticidal chemicals is diminishing rapidly. To a lesser extent the range of chemicals approved for use on minor crops is also diminishing as a result of more stringent and expensive testing. To counter these situations intense activity has been aimed at the introduction into crops of resistance/tolerance traits from a diverse range of sources. For almost a decade virus-derived sequences have been expressed in crop plants (Sanford and Johnston, 1985; Harrison, 1992; Lindbo and Dougherty, 1992a,b; Fitchen and Beachey, 1993) and there have been numerous regulated field trials. Since these tests have proceeded in many parts of the world without producing evidence of unexpected harm, there is now a strong industry push towards commercial use. Indeed, unregulated use in the USA of squash (*Cucurbita pepo* subsp. *ovifera*) containing two potyvirus-derived capsid coding sequences was authorized in 1994 (APHIS/USDA Petition 92-204-01 for determination of non-regulated status of ZW-20 squash). The possibility of undesirable consequences following dissemination of virus-derived genes or sequences from virus-like hyperparasites (e.g. satellites; Liu and Cooper, 1994) is the subject of considerable debate and there is no consensus. Importantly, insofar as it is possible to know opinion, the lay public is not at present convinced of the value or the need for the release of genetically engineered viruses into the environment. Distrust and frustration among a (probably) tiny part of the lay community has overspilled into hostility and, occasionally, even violence.

Transformation (genetic change) in plants has been effected in numerous ways (Potkyrus, 1991) but is routinely achieved by the integration of a plasmid (from the root parasitic bacterium, *Agrobacterium tumefaciens*) containing sequences from a pararetrovirus (cauliflower mosaic caulimovirus). In the absence of selective advantage, the extraneous DNA is more or less rapidly eliminated because pararetroviruses use the host nucleus to transcribe viral DNA into RNA and the cytoplasm for reverse transcription of the RNA into DNA again. However, the integration is normally monitored by linkage to an antibiotic- or herbicide-resistant phenotype. In the context of transformation in vertebrate cells, papovaviruses and papilloma viruses were assessed initially but, because these genomes resemble caulimoviruses in that they exist as episomes in the cytoplasm of their hosts, they were rapidly superseded. Viruses containing ssDNA genomes such as geminiviruses resemble polyoma viruses of vertebrates in their expression characteristics and are alternative gene vectors, albeit with very modest payload, which have been engineered to shuttle between bacteria (facilitating genetic manipulation) and plant cells (for expression).

Building upon experience of live adenovirus vaccination programmes in military populations, laboratory-made recombinant adenoviruses have been used to express a variety of immunogens – notably glycoproteins from lentiviruses, hepadnaviruses or rhabdoviruses. The same convenience and abilities to express glycoproteins are properties possessed by pathogenic viruses from invertebrates (notably a baculovirus from the insect *Autographa californica*); this system (King and Possee, 1992) became routine even before the complete molecular biology/genome organization of the virus was known (Ayres *et al.*, 1994). However, poxviruses (which have no size constraint on packaging) have now become the standard laboratory expression system for use in the context of human/veterinary medicine; they have also been taken into the environment – for the dispersal of rabies vaccine in chicken heads targeted at fox populations in Europe (Moss, 1991; Baxby and Paoletti, 1992).

dsDNA systems replicate with fidelity whereas RNA-based systems are more ephemeral. Nevertheless, picornaviruses are being assessed as vectors for the delivery of peptide-based vaccine systems and chimeral picornaviruses, including those which infect plants, are potentially useful in the development of diagnostic reagents or

pharming (use of plants to produce fine chemicals of pharmaceutical value) if not necessarily direct prophylaxis. Infectious transcripts have been produced from DNA copies of numerous viruses and also satellite RNAs, thereby facilitating the possible use of plus-stranded RNA viruses as vectors. Furthermore, heterologous gene expression has been reported following sequence substitution in the capsid coding sequence of bromoviruses or tobamoviruses (Joshi *et al.*, 1990; Ahlquist and Pacha, 1991).

For the RNA-based vectors to replicate and spread independently within plants there is a need to maintain the genetic factors for replication, encapsidation and movement. In an effort to constrain opportunities for the dissemination of RNA-based gene vectors, one or more of the essential functions have been deleted and used by allowing different viruses replicating together to 'feed' one another. Although production and use of genetically modified viruses in licensed laboratories is routine, recombination (Chang *et al.*, 1988; Lai, 1992) is an omnipresent consideration. New functional virus genomes may arise from recombination involving disarmed vectors, co-infecting helper viruses (de Jong and Ahlquist, 1992) or transgenic genes (Gal *et al.*, 1992; Greene and Allison, 1994). Such processes could complement sequence deficiencies in superinfecting viruses and, before deliberately releasing genetically engineered viruses into the field, the likelihood of any such hazard needs to be critically assessed on a case-by-case basis because viruses differ in their intrinsic proness for such interactions.

The simple act of transforming a plant or animal so that it thereafter expresses a virus-derived product, may not necessarily have much impact on the ecology of any ecosystem into which the transformed object is deliberately introduced. However, the likelihood of uncommon events leading, at an extreme, to harm is presumed to be scale-dependent and the more transformed plants in a community, the greater the prospect of unexpected results. Viruses are mutable and viruses with RNA genomes are especially prone to unpredictable change. Even point mutations have been implicated with enhanced pathogenicity (Sleat and Palukaitis, 1990; Chen *et al.*, 1994). Recombination with host sequences has also been correlated with a new pathotype (Kuwata *et al.*, 1991). Nevertheless, transgenic plants containing satellite-derived sequences alone or in combination with capsid coding sequences have been widely planted, particularly in the Peoples Republic of China (Tien and Wu, 1991).

Most point mutations, deletions and gene rearrangements in viruses were presumed to be lethal, or at least to result in the withdrawal of the new genotypes from the pre-existing viral gene pool. However, this view is probably naive, particularly having regard to rampant recombination which has been recognized in some taxa (e.g. luteoviruses (Mayo and Jolly, 1991; Gibbs and Cooper, 1995). Retrospective assessments based on analyses of extant sequences indicate that genes have been exchanged between viruses in different taxa, with different current vectors and, insofar as is known, with distinct natural host ranges. This indicates a hitherto unrecognized diversity and gene pool size. Even though transgenic genes will be refined to minimize problems which can be attributed to specific sequences, the emergence of new gene combinations with unpredictable epidemiological potential can be envisaged – particularly if recombination occurs in context with transcapsidation. On the basis of very incomplete knowledge (visual surveys for severe disease in a tiny number of crop plants and their relatives) the view has developed that most plant viruses are constrained within a small number of species. However, the fault in such a simplified concept of safety based on host restriction is thrown into question by the commonplace experience of new epidemics in long-established crops and the subsequent discovery (when closing stable doors after horses have bolted) that considerable genetic and phenotypic diversity exists even within one virus taxon (e.g. a serotype). Judgements about virus distribution depend upon the best available detection systems, but all of these methods have flaws. Serology is quick and convenient and, until polymerase chain amplification of nucleotide sequences (Mullis, 1990) became routine, was – for reasons outlined by Cooper and Edwards (1986) – the method of first choice in surveys. However, serology reflects external features (e.g. structural capsid proteins) and thereby measures the properties deriving from only a small part of the viral genome.

Despite constraints imposed by interference and vector preferences, multiple infection in plants is not uncommon (Abdalla et al., 1985); estimates of occurrence and the identities of the viruses present are limited by the rigour of the tests used. It remains to be seen whether there will be ecological consequences for non-transgenic plants in any wild community but it is inevitable that opportunities for gene exchange will increase over that in nature when all plants in a crop have the virus-derived attribute rather than the minority which is presumed to be the current (background) situation. In reality, no substantial surveys have been done to address the prevalence of resistance genes in naturally regenerating periagricultural environments. However, using as an index the frequency of success recorded by traditional plant breeders seeking natural resistance/tolerance genes in wild relatives of commercial crops, these traits seem to be very rare or absent (Taylor and Ghabriel, 1986; Larkin et al., 1989). Numerical data relating to field exposure of wild plants to viruses are not widely reported and data concerning, for example barley yellow dwarf resistance in close relatives of cereals might not reflect the situation in other grasses. Nevertheless, commercial field crops containing virus-derived transgenic genes are (will be) substantially different from natural resistance/tolerance traits – both in terms of their ubiquity and (presumably) their distinctive chemical characteristics.

Viruses in aquatic environments

5

5.1 INTRODUCTION

Vast numbers of animals, birds, reptiles, insects, etc. require intermittent access to water and undoubtedly shed viruses while drinking, swimming, etc. However, little is known about the range of viruses liberated and their fate in water. This is partly attributable to the technical difficulty of isolating viruses from large volumes of water liable to have a heavy load of suspended solids and to the specificity of virus-detection methods which are also insensitive (Primrose *et al.*, 1981). Most is known about a few systems which have been studied in relation to human disease. Indeed, the study of viruses in water has to a large extent been justified because ground water supplies (e.g. artesian wells), which were formerly adequate for urban man, are becoming insufficient to satisfy the requirement for safe drinking (potable) water. Urban/metropolitan communities now need to exploit surface waters (e.g. rivers) for drinking, and it has long been known that these are liable to be contaminated, not only with human faeces but also with plant pathogenic agents (Koenig, 1986). In scattered rural communities the risks of virus infection from water seem less for the human population, but there may be important unexplored consequences for the natural assemblages of animals that routinely pollute with faeces and urine their own, sometimes limited, water supplies. Because fresh and salt water are the media in which diverse elements of the microflora and microfauna exist, it is likely that their attendant viruses will have been selected to survive in these aqueous environments. Figure 5.1 illustrates some routes of virus transmission in water.

5.2 INDIGENOUS SOURCES OF VIRUSES

Being the common denominator of life, water in virtually all natural states has a very diverse flora and fauna, much of it microscopic. Little

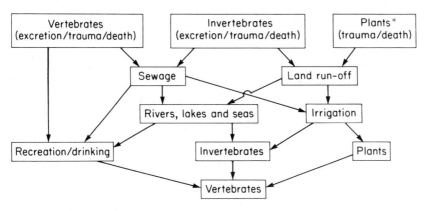

Figure 5.1 Some routes of virus transmission in water. *Bacteria, algae, fungi and higher plants.

is known about viruses that infect most of this biotia but it would be very surprising if they did not collectively host multitudes. A few scattered data are, however, available and indicative. Viruses (cyanophage; Padan and Shilo, 1973) pathogenic for blue-green algae consistently infest streams, fishponds and reservoirs in the USA, former USSR, UK, Israel and India and are known to exist in soil water. Hence it would be surprising if the diversity of virus-like agents (Brown, 1972; van Etten et al., 1991; Muller and Stache, 1992) recognized in algae (which constitutes some 50% of the world's biomass) did not also survive in water. Probably, marine mammals such as sealions, fur seals, elephant seals and perhaps also whales liberate caliciviruses in their faeces, as do pigs and mink fed on calicivirus-infected sealmeat (Wilder and Dardiri, 1978). An additional range of viruses is probably liberated by amphibians (e.g. a herpesvirus of frogs which is shed in urine), and some of the diverse array of virus-like agents that have been isolated from diseased fish (Hill, 1981), such as the rhabdovirus associated with infectious haematopoietic necrosis of sockeye salmon, are excreted in urine and faeces. Similarly, the virus associated with infectious pancreatic necrosis in freshwater trout is liberated into water with sperm and eggs as well as urine. Doubtless, parasites and pests (e.g. protozoa, nematodes, fungi, annelids and arthropods) will be found as essential vectors of some fish viruses but several have been transmitted experimentally when gills of infectible fish were brushed with virus-containing extracts of infected specimens. Plausibly viruses in water are, for fish and perhaps other aquatic animals, the equivalent of viruses dispersed in aerosols that inoculate terrestrial vertebrates, although some fish in marine environments drink water to maintain their osmotic balance and thereby enhance their exposure to viruses.

5.3 SOME EXTRANEOUS SOURCES OF VIRUS CONTAMINATION

The significance of natural virus release from flowering plants is mentioned elsewhere (p. 56) in relation to virus acquisition and transmission by chytrid fungi. Most viruses pathogenic for plants and many pathogenic for invertebrates can survive for weeks or months in water and it is noteworthy that any viroids entering soil water from infected plants would probably be almost as resilient. The extent and importance of this contamination seems slight but it has not been assessed critically. By contrast, there is considerable evidence implicating water as a medium for the transmission of viruses pathogenic for vertebrates (Berg, 1967). As mentioned in Chapter 2, ocular, respiratory and genital shedding of viruses is known to occur and from time to time undoubtedly contributes inocula into water. Faeces seem to be particularly important sources of viruses that contaminate water; dozens of distinct viruses are discharged in this way. Of these, adenoviruses, rotaviruses, reoviruses (Flewett and Woode, 1978) and enteric picornaviruses are among the most resilient, but a range of parvoviruses recognized chiefly in rodents, cats, dogs and birds are similarly stable. All remain viable for long periods in water. Faecal shedding of human adenoviruses is especially common in children and discharge into swimming pools seems to explain some cases of conjunctivitis and/or pharyngitis that develop after immersion. Similarly, epidemic diarrhoea in adults and acute infantile gastroenteritis are associated with rotaviruses which are liberated with faeces in massive amounts and are suspected to be water-borne routinely (Lycke et al., 1978).

Sewage from human settlements contains faeces, blood and urine from diverse non-human sources, such as pets which are almost as numerous as the human population, the ratio of dogs to people being 1:6 in France and 1:25 in West Germany (Table 5.1). There are estimated to be more than 30 million dogs in the USA plus a similar number of cats, 15 million cage birds, 600 million fish and 10 million other pets (primates, amphibians and rodents). Water-borne inoculum probably explains some of the recoveries of human enteroviruses from rabbits, canines and budgerigars but transmission in the opposite direction is also possible. Roughly 20 000 tonnes of canine faeces are deposited annually on the sidewalks of New York, an index of the contamination in all metropolitan areas. The human population is undoubtedly exposed to massive amounts of virus shed in faeces or in other ways from pet animals and commensal rodents (Ackerman et al., 1964).

In rural areas the death of infected animals may from time to time liberate viruses into surface waters but wild rodents and birds seem

Table 5.1 Indicators of the proportion of pets to people

Country		Pet population (millions)	Human population (millions)	Ratio
France	Dog	7.5	51	1:6
	Cat	7.1		
UK	Dog	5.8	55	1:10
	Cat	4.5		
USA	Dog	33–35.4	200	1:6
	Cat	33.6		
West Germany	Dog	2.4	61	1:25
	Cat	2.3		

more notable and important sources of virus pollution. The cloacae of waterfowl squirts out and sucks in water which is contaminated with faecal microorganisms and viruses as a consequence. Numerous avian influenza A viruses have been found in faeces and also in lakes where feral as well as domesticated ducks congregate, thereby providing opportunities for the infection of avian as well as mammalian species (Webster *et al.*, 1978). Surface waters are also liable to be contaminated with viruses from faeces of domesticated livestock (Derbyshire and Brown, 1978); porcine and bovine enteroviruses have been isolated from faecal slurry and, in avian faeces, virus-like particles resembling rotaviruses and enteroviruses have been detected by electron microscopy. Additionally, vaccine-derived picornaviruses associated with, for example, foot-and-mouth or swine vesicular diseases probably find their way into surface waters in countries where control of these diseases by vaccination is preferred to slaughter. In the UK, stringent precautions are taken to eliminate the prospect of foot-and-mouth disease virus entering water supplies after deep burial of animals slaughtered for control.

5.4 THE FATE OF VIRUS-CONTAMINATED SEWAGE

Because of minimal cost, discharge of sewage into fresh or sea-water is commonplace, almost irrespective of the magnitude of the human component in the sewage catchment zone. To indicate the scale of the problem, about 90% of Canada's urban population has sewage collection, but the effluent from 40% of this population is never treated. The river Rhine receives more than 70% of the total sewage of West Germany, and, in less populated areas, enteroviruses have been detected 25 km downstream from a single known source of sewage. In the USA, where 3–18% of drinking water is derived from surface waters, 5% of sewage is discharged directly into waterways and a further 22% follows the same route after a minimal treatment (settling and microbial action in cesspools). The balance of sewage is treated (Anon, 1979a) before discharge (Figure 5.2).

The amounts of aesthetically displeasing solids on which viruses are liable to be adsorbed but still infectious are lessened by filtration and biological processes mediated by a diversity of unspecified bacteria, fungi, algae, protozoa and other invertebrates. This (secondary) treatment is applied in a variety of ways which depend upon oxygen generated by algal photosynthesis facilitating heterotrophic degradation by bacteria and fungi while numerous anaerobic processes operate in parallel. Few treatments applied to sewage are wholly effective in removing viruses (Anon, 1979b; Bitton, 1980). Indeed, reflecting host occurrence, cyanophages and bacteriophages are commonplace during and after sewage treatment. Similarly, the amounts of enteroviruses (largely the vaccine-derived poliomyelitis viruses) recovered in the primary effluent routinely range between 10 and 100% of that in the influent sewage. Bubbling and splashing at various stages in sewage treatment inevitably liberate aerosols that are potential sources of inocula to animals in the vicinity. Viruses liable to be scattered in this way include enteroviruses and adenoviruses which survive best in air at high relative humidity, and myxoviruses/paramyxoviruses which remain infectious longest at low relative humidity. However, there is little evidence that occupational exposure to sewage enhances the risk of humans showing signs of diseases due to viruses (Clark *et al.*, 1976). No tests seem to have been done on rodents, birds and other wildlife that frequent sewage-treatment plants.

5.4.1 AGRICULTURAL USE OF WASTE WATER

Sewage treatment has two main products: waste water and sludge which is semi-solid biological and inorganic matter having diverse origins. The most convenient disposal routes for both are into land or water.

Sewage is a traditional source of nutriment for soils and crops and in recent years slurry from intensively reared pigs, poultry, etc. or waste water effluent from sewage treatment works has been used similarly. There is little concrete evidence of consequential risk to humans, their animals (or indeed their crops) except insofar as inorganic pollutants (e.g. zinc) and radionuclides of strontium, caesium and other rare earths are recognized as potentially harmful (Baldwin *et al.*, 1976). Nevertheless, vegetables that are eaten raw after sewage effluent has been applied to them can be a source of enteric viruses such as cause poliomyelitis (Larkin *et al.*, 1976). Washing of vegetables (in potable

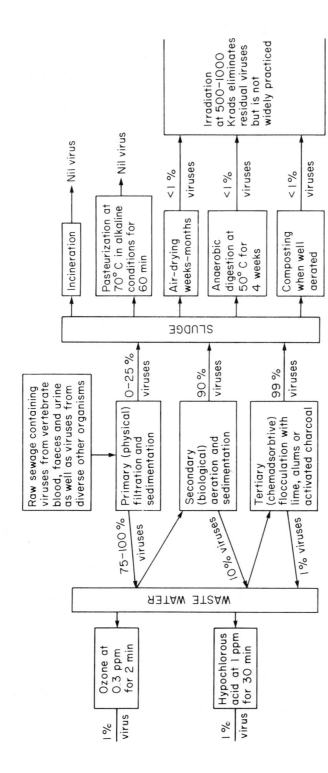

Figure 5.2 Flow chart showing efficiency of virus removal at different stages in sewage treatment.

water) does not remove all viruses that are adsorbed to foliar surfaces and the potential threat is recognized by many states in the developed world which prohibit the use of sewage effluent on plants destined for human consumption. Fortunately, surface contamination of plants by viruses seems to be short-lived (viruses applied are not normally detected after 1 month*). Indeed, plants seem to inactivate viruses on their surfaces at rates greater than can be attributed to the effects of solar radiation, desiccation or rainfall and tannins in plant juices have broad-spectrum antiviral activity (Konowalchuk and Spiers, 1976). Intriguingly, there are a few published data implying that viruses pathogenic for vertebrates are translocated within the vascular system of plants (Murphy and Syverton, 1958; Murphy et al., 1958). Mouse encephalomyelitis virus was detected in the stems and leaves of peas, potatoes and tomatoes with their roots in water containing the virus and other parts protected from splash contamination. It is difficult to judge the importance of this finding (which does not seem to have been re-assessed experimentally) but agricultural use of waste water is strongly suspected to present a health risk from aerosols generated during application. Enteroviruses have been detected in air 50 m down-wind from sprinklers, and, in an agricultural commune in Israel, Katzenelson et al. (1976) noticed a higher incidence of virus hepatitis and influenza during summer (when oxidation pond effluents were used for irrigation) than during winter when such irrigation was not used.

5.4.2 VIRUSES IN DRINKING WATER

Depending on the system of application, sewage effluents may move downwards into the soil or laterally as run-off which finds its way into surface rivers, lakes, etc. where it is liable to be re-used for irrigation or otherwise. En route, viruses tend to be adsorbed to colloidal matrices but the amounts sequestered in this way vary enormously depending on soil structure, composition, pH and the rate of downward flow. Furthermore, the viruses are not inactivated. Poliomyelitis virus retains its infectivity even when adsorbed to soil, and, depending on the temperature, moisture content and presumably the indigenous flora and fauna, persists for weeks to months. Vertebrate viruses have been detected in wells up to 7 m deep when heavy rain followed the application of slurry/waste water to soils. Because the factors that influence virus movement and survival in soil are not fully understood, great caution is needed when using waste water for irrigation near potential supplies of drinking water. Water for drinking that is com-

*Polymerase chain amplification does not necessarily indicate biological activity.

monly abstracted from reservoirs fed by surface rivers but may be derived directly from waste water is normally disinfected with halogens after filtration. Chlorine that has been used traditionally is not always adequate. Enteroviruses, and to a lesser extent reoviruses and adenoviruses, are much more resistant to chlorine than coliform bacteria against which the process was directed originally. A spectacular outbreak of viral hepatitis type A occurred in 1955 when about 30 000 people were infected because the piped water supply in New Delhi, India, became contaminated with sewage during a flood (Viswanathan, 1957). Even though the danger was recognized and chlorination was applied to achieve a final concentration greater than 2 ppm available chlorine, which was adequate to prevent epidemic poliomyelitis and bacterial dysentery/typhoid, the treatment did not eliminate the enterovirus-like agent of hepatitis A. Alternative treatments, such as iodine, bromine or ozone (above a critical exposure time/threshold concentration), or ultraviolet light (200–300 nm) in the presence of photosensitizing dyes, seem more effective against viruses liable to contaminate drinking water. Unfortunately, because these processes are somewhat more technically exacting to apply than chlorination, they are not used routinely on a large scale. Chang (1968) suggested that a (virologically) safe potable water could be produced from polluted sources if the efficiency of virus removal was greater than 99.993%. A water-treatment system with the required efficiency (99.995%), due largely to flocculation and chlorination, has been operated on a pilot scale (Guy et al., 1977) but costs are substantial.

5.4.3 THE FATE OF SEWAGE SLUDGE

Sludge is a malodorous mixture that contains toxic amounts of nitrates, detergents and metals as well as vertebrate pathogens including viruses. Consequently, direct disposal of sludge into land or water can result in unacceptable pollution with this broad spectrum of materials. Nevertheless, sludge has been used as a soil supplement under short-rotation coppiced trees such as poplar or willow. To lessen the risk, sludge may be treated in one or more of the ways listed in Figure 5.2, but the treatments are not equally effective in removing viruses. Vertebrate pathogens are virtually eliminated when sludge is held at a temperature of 70°C for 1 hour before disposal but this process, like incineration, is very costly in terms of energy consumed. Moreover, incineration presents additional problems attributable to smoke and smell. A less costly alternative treatment produces an odourless compost from sludge that has been mixed with woodshavings and allowed to stand in heaps for a month or so. During this time the heaps are aerated and indigenous microbes generate heat (in the range of 50–70°C) which tends to eliminate enteroviruses but has a lesser effect on reoviruses (Ward and

Ashley, 1978). Sludge disposal by injection below the soil surface minimizes problems attributable to aerosols, surface run-off and smell, and may ultimately supersede incorporation of sludge by ploughing, which is more widely practised.

5.4.4 THE CONTAMINATION OF SEA WATER

Because of the hazard to marine life and the concern for human health, the practice of discarding sewage and sludge into sea water is diminishing but continues to be prevalent (Gameson, 1975). One study in Hawaii estimated 8×10^{10} enterovirus particles were discharged into ocean water every day, and in the Mediterranean, where specific action to lessen the contamination has been proposed (Brisou, 1976), a range of enteroviruses has been detected up to 1.5 km from the nearest known discharge point. Off the shore of Texas, in Galveston Bay, enteric viruses were recognized as far as 13 km from the known source of faecal contamination, although the discharge of untreated sewage from ships is commonplace and may explain the observation. Fortunately, unknown biological factors in sea water cause virus concentrations to decline at rates determined by temperature, during days at 22°C and months at 3–5°C. Nevertheless, there is a well-established risk to swimmers because bathers ingest 10–50 ml of water per bathing period. An additional hazard is attributable to viruses in the sea being transferred into the atmosphere. Surf produces 3×10^5 bubbles/m² per second and Baylor et al. (1977) showed that the concentration of viruses in the air can be enhanced over that in the main body of sea water as a result of virus particles adsorbing to bubbles that rise through the water. This phenomenon has been used in the laboratory to concentrate the virus of foot-and-mouth disease (Morrow, 1969). Predictably, some virions adsorb to particles liable to sediment and some of this material enters food-webs. Bivalve molluscs which feed on particles of organic matter (liable to have virions adsorbed) filter water at rates estimated between 4 and 20 litres/hour, depending on factors such as the species of mollusc, temperature, pH, salinity, etc. Removal of 70% of applied enteroviruses by shellfish has been claimed: numerous enteroviruses as well as coliphages and reoviruses have been isolated from shellfish (Gerba and Goyal, 1978). Not all the virus particles enter the digestive tract (and those that do are not certain to be rendered uninfectious as a consequence); some adhere to mucous-coated outer surfaces where they may be retained. Predictably, shellfish that are cultured in waste water-rich environments near shores (and are liable to be eaten uncooked) have been associated with hepatitis and gastroenteritis, sometimes with high human morbidity (Dienstag et al., 1976). Furthermore, crustacea such as crabs and other detritivors (e.g. polychaete worms) are capable of contamination with enteroviruses acquired from shellfish. Consequently, at later stages in the food-chains, racoons in freshwater and otters in marine situations as well as birds in littoral environments may be as much at risk from virus-contaminated invertebrates as the human population. The marine environment has been identified as a potentially important source of caliciviruses which infect sea mammals and contaminate (perhaps also infect) fish and nematodes and cause disease in pigs on land (Smith *et al.*, 1978).

Viruses undoubtedly contaminate natural waters, which can serve as vehicles of transmission in specific circumstances, but the importance of this route in the total epidemiology of most viruses is open to question. Because of commercial exploitation as human food, fish are a form of wildlife that is closely monitored for disease and population change. A few viruses associated with gross abnormalities are known but effects on populations are little studied (when infected animals die at sea, there is little record) and usually trivialized because of ignorance. It took a dramatic epizootic in marine mammals to focus attention on the occurrence of viruses in marine mammals. Canine distemper sprang to new prominence in 1988 when this virus – long associated with death and debilitation of young canines – was presumed to be a cause of seal death in coastal regions of the Baltic and North Seas. Seals in land-locked Lake Baikal almost certainly died from canine distemper. However, the other seal viruses are serologically distinctive; phocid morbilliviruses have been isolated from a variety of marine species and serological surveys suggest endemicity in dolphins, porpoises, etc.

Even though the dissemination and survival of viruses in the aqueous environment is now investigated in a little more detail than hitherto, reliable information is very sparse. Coliform bacteria which have been traditional indicators of faecal pollution, are inadequate guides to the possible human health risk from viruses, and alternative standards of water quality based on testing for enteric viruses undoubtedly – and urgently – need to be developed and evaluated.

Strategies of virus maintenance in communities

6

6.1 INTRODUCTION

Knowledge is rapidly accumulating about how viruses perennate within individuals. By contrast, much less is known about mechanisms facilitating virus maintenance within populations or communities. Even though perennation and maintenance phenomena are often intimately related, there is an understandable tendency for the overt disease to attract more attention than the silent infection that might have been its source. Virological knowledge has developed in depth rather than breadth as a series of largely independent studies with perspectives circumscribed by the properties of the differing target populations. In this context target populations are the primary subjects of investigation in relation to a specific virus. To some extent, the approach is due to the emphasis placed by urban Man on those few viruses (e.g. measles, Paramyxoviridae) for which humans seem the most important natural host (Figure 6.1(a)). However, viruses circulating within only one population seem exceptional. Of the positions which a target population can occupy in relation to the biology of a virus, the most common seems to be one in which it only has a share in the maintenance, other populations grouped under the term 'wildlife' being more or less important, as with monkeys vis-à-vis humans in sylvatic yellow fever (Togaviridae) or the potyvirus watermelon mosaic (Adlerz, in Thresh, 1981) which alternates between wild and cultivated (target) cucurbits (Figure 6.1(b)). In many instances the target population is not involved in virus maintenance, e.g. rabies (Rhabdoviridae) and arenavirus in humans (Casals, in Evans, 1976), maize rough dwarf (Reoviridae) having wild grass hosts (Harpaz, 1972) but damaging maize (Figure 6.1(c)).

From our patchy knowledge of viruses in the natural environment, it has often seemed reasonable to assume that spatial or temporal heterogeneity adequately explained the coexistence of viruses in the

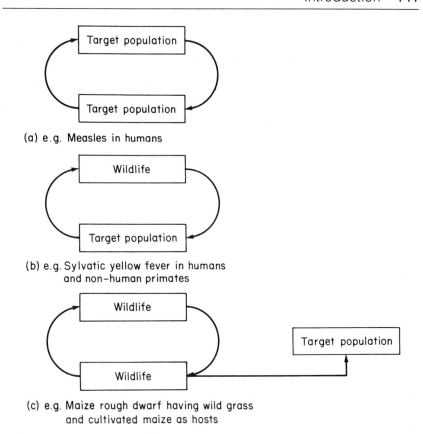

(a) e.g. Measles in humans

(b) e.g. Sylvatic yellow fever in humans
      and non-human primates

(c) e.g. Maize rough dwarf having wild grass
      and cultivated maize as hosts

Figure 6.1 Three options for the circulation of viruses within or between target populations and others grouped under 'wildlife'.

vicinity of target populations. An alternative approach has explained the experiences in terms of host population dynamics. Viruses need a constant supply of infectibles. The only exceptions are when viruses have effective methods of perennation (e.g. vertical transmission) rendering horizontal transmission irrelevant or when they have significant potential for independent survival outside the cells of a host. Although the latter property is unusual, adsorption to fomites and soil particles probably enhances to an important extent the survival prospects of numerous viruses (cf. Allen, 1981). In any event, virus maintenance ultimately depends on the rates of population turnover in the virus being in step with the rates of turnover of hosts and, when appropriate, vectors. The virus has a regeneration time that is determined in part by the time taken for one newly infected cell to become a source of infection for another, but also by other temporal events such as acquisition access time, the latent period in vectors, the journey time from infector to infectibles, etc. Although the concentration of viruses in systemically invaded plants tends to diminish, the viruses do not seem to be eliminated from plant hosts, as tends to happen with vertebrates in which antibodies circulate. In this respect viruses in plants resemble herpesviruses and others which may exist within their vertebrate hosts while being inaccessible to the components of the immune system. Thus, although the duration of a host's infectiousness in relation to host regeneration time is in general critically important, for perennial plants (e.g. trees) maintenance is possible with only very occasional opportunities for dissemination.

In the absence of detailed knowledge about how the target populations perpetuate themselves, it is difficult to be definite but it seems likely that virus growth often occurs in cells that are genetically as well as spatially isolated from the regenerative core of the cell (or multicellular) populations. When the detailed information is available (e.g. for humans) it is realistic to deduce that viruses killing hosts which have passed reproductive age (e.g. human females over 50 years of age) are unlikely to have much impact on the survival potential of the host population. Plausibly, factors contributing to the elimination of a potential host away from its natural environment are unimportant to the maintenance of the host population as a whole. Thus, bacterial pathogens of humans that are normally transmitted from hand-to-mouth or by flies from faeces to food (e.g. *Shigella* spp. causing acute dysentery) are liable to enter soil/surface water in which they can grow but rapidly succumb in competition with other organisms. The elimination of *Shigella* spp. from river water by viruses is unlikely to diminish their survival as intestinal parasites, except when ingestion of contaminated water is a significant source of inoculum.

Drawing upon the diversity of experience with viruses, some of the similarities and differences between virus–host systems are outlined below in the hope that speculation and comparison will stimulate experiment and further hypotheses, resulting in a better appreciation of viruses within a broad biological canvas.

6.2 AGRICULTURAL PLANT POPULATIONS

Agricultural/horticultural crops have been notable subjects of virological study but with few exceptions they offer no prospect of virus maintenance. Although crop plants are more or less genetically uniform and individuals are often phenotypically indistinguishable, it is worth noting that they are not necessarily equally infectible by viruses (Louie *et al.*, 1976). Nevertheless, analysis of the spatial and temporal progress of viruses in these systems is facilitated because the infectives in the target population are static. The observed gradients of disease incidence

(Figure 6.2) have given valuable insight into the habits of virus vectors, and factors influencing their movement have been described in biological (Thresh, 1976) and mathematical terms (van der Plank, 1960).

With few exceptions, the sigmoidal increases in disease incidence with time (Figure 6.2(a,b)) involve components of spread both within the target population and from outside. Even though these two components have not been separated adequately in most instances, the accumulated knowledge has predictive value. Analyses have provided useful guidelines for virus–disease management, prospects for containment and possibilities for manipulating transmission dynamics through use of insecticides and in other ways. However, in agriculture, the population of infectibles is fixed by the original planting density, and natural regeneration is positively discouraged. Consequently, this

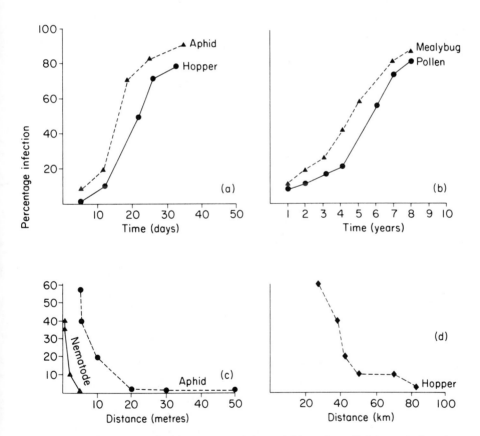

Figure 6.2 Diagrams showing temporal (a and b) and spatial (c and d) patterns of disease incidence associated with viruses dispersed by aphids, hoppers, mealybugs, nematodes or pollen. Redrawn after Thresh (1980b) with permission.

management creates a system in which the incidence of disease (reflecting virus infections) is directly proportional to the size of the target population modified by factors influencing the transmission efficiency, e.g. plant spacing (Figure 6.3), location in relation to sources of infection (Figure 6.2(c,d)), effects of weather on vectors and on properties of the virus-replication cycle within cells. There is some loss of infectibles due to natural causes and an additional loss (at a rate of α individuals in unit time) directly attributable to the virus but the most significant change is due to the periodic harvest of the residual crop and the destruction of the survivors (Figure 6.4).

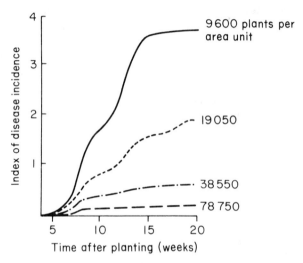

Figure 6.3 Effect of planting density on the percentage of groundnut plants infected with rosette virus. (From A'Brook, 1964.)

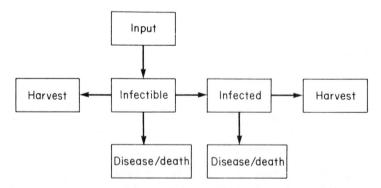

Figure 6.4 Diagram showing the flow of components of a target population (food/fibre plants) challenged by a virus.

6.3 NATURALLY REGENERATING POPULATIONS

In natural systems, the rate of input of infectibles is not some arbitrary constant but is set by the various regeneration and depletion processes that operate. The infectible population regenerates infectibles at a rate of b individuals in unit time, and those infected individuals that survive similarly contribute to the pool of infectibles entering the system. Additionally, both the infectible and the infected populations are likely to contribute a new class of genetically immune individuals which are withdrawn from the system until the virus changes its host range. The rate of population conversion to genetically immune has consequences for the maintenance of virus because space and other resources are limited and there is competition between the new and the old populations. However, the rate of this conversion is likely to be slow in higher plant and animal populations. By contrast, because bacteria have rapid rates of reproduction, the consequences of competition may be obvious as complete population conversion during an interval that is significant when compared with the regeneration period of the virus.

In this situation (Figure 6.5) the infected population which contains Y individuals in time t may not compete so effectively with healthy infectibles, thereby enhancing the rate of death (d). In addition, the infected population need not contribute to the input (B) of infectibles at the same *per capita* rate (b) as do healthy infectibles, i.e. b' need not be the same as b (when the virus is vertically transmitted in pollen, sperm, etc. this has the same effect).

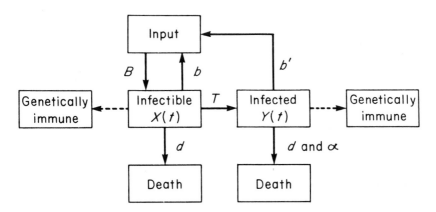

Figure 6.5 Diagram showing the flow of components of a naturally regenerating population presumed to lack post-infection responses (e.g. antibodies, complements, interferons) affecting the rate and extent of invasion when challenged with a virus.

6.4 VERTEBRATE POPULATIONS

Within populations of vertebrate hosts, an additional factor becomes important. Vertebrates that survive infection (at a recovery rate of R individuals in unit time) are liable to exhibit an acquired immunity. Whereas the genetically immune components of other populations could be discounted, this is not possible in this instance. Immune responsiveness is a complex variable but at its simplest might be assumed to exhibit a potential for reversion that converts immunes to infectibles at a rate of γ individuals per unit time. Additionally, that part of the population with acquired immunity (Z individuals in unit time) is liable to die at the natural rate (d) while contributing to the input of infectibles at a rate of b'' individuals per unit time (Figure 6.6).

6.5 APPLICATION OF PREDATOR–PREY MODELS TO VIRUS–HOST DYNAMICS

Theory concerning the habits of ecologically interacting species living in the same volume has been largely based on interpretation of the dynamics of real invertebrate populations on the one hand or notional populations on the other. In the latter, no attempt was made to re-present each aspect of diversity of the real world. However, the initial simplified theoretical framework can evolve to accommodate increasing amounts of data. In recent years, theoretical models used by ecologists

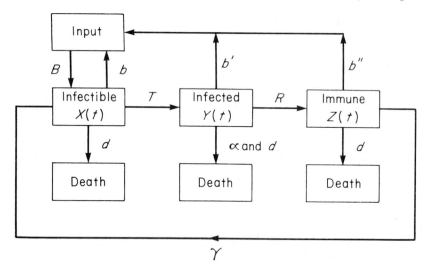

Figure 6.6 Diagram showing the flow of components of a naturally regenerating population possibly having post-infection responses (e.g. antibodies, complements, interferons, etc.) affecting the rate and extent of invasion when challenged with a virus. (After Anderson and May, 1979.)

to explain and predict the population fluctuations which characterize interactions between predators and prey have been applied to systems involving viruses and animal or bacterial hosts and these have revealed some common principles. Models used by ecologists to describe predator−prey interactions consider idealized uniformity both of habitat and species; the latter interact at random with frequencies determined solely by their relative concentrations. Consequently, the models are most directly applicable to viruses spreading by contagion although Murray and others (1986) predicted that rabies introduced in the fox population in the UK had epizootic character until the density of infectibles diminished to a value that was too low to allow transmission. However, once the fox population built up, epidemic spread followed.

The additional complexity attributable to vectors is, however, manageable within a similar framework. Interestingly, although the models have solutions predicting stable equilibria or stable limit cycles of oscillation, classical experiments on predator−prey interactions such as those of Gause (1934) did not always exhibit stable states of coexistence in uniform habitats. Irrespective of initial density of the two populations, predators can eliminate prey, then starvation eliminates the predators. In the context of virus biology, predator−prey theory has been applied to a few experimental systems. The most notable experimental data were derived from studies on mouse poxvirus in laboratory rodent populations where infection occurred largely as a result of inhalation and resulted in about 50% mortality. Despite this high lethality, a mouse colony initially containing 25 healthy and 20 virus-infected animals (from which dead mice were removed and to which three infectible mice were added daily for a 30-month period) increased to a constant level of 230. Furthermore, the virus was maintained and, at equilibrium, approximately one-fifth of the animals were infected, one-fifth were healthy but infectible and the remainder were immune. Although the smallest population facilitating maintenance of mouse pox was not determined in these experiments (Greenwood et al., 1936), similar studies by Fenner (1948) subsequently showed that maintenance of mouse pox was possible when the mouse population was 70. This series of observations, which were based on instantaneous input and output (without the time delays implicit in natural regeneration processes), were analysed by Anderson and May (1979) in terms of equations describing the various rates. Using the notation applied to the compartment model (Figure 6.5), they calculated that the equations had a stable equilibrium solution facilitating virus maintenance within the populations only when,

$$B/d > \frac{(\alpha + d + R)}{T}$$

In other circumstances, the model predicted that virus would be eliminated and that the infectible population would have a new equilibrium with the total number of animals being the input, modified only by the natural death rate. Theory and the laboratory experience with mouse pox were in agreement. The equation deduced by Anderson and May is a generalized expression of the phenomenon recognized by Black (1966) and others that a critical human density (determined by natural birth or immigration) must be exceeded for maintenance of a virus such as measles (Figure 6.7). The input rate which was constant in the mouse–mouse pox experiment (but would be dynamic and possibly responsive to infection in the real world) is an additional variable that can be accommodated within the scheme outlined above to assess the potential of viruses to control natural populations.

6.6 REGULATION OF POPULATION GROWTH BY VIRUSES

There is a mass of circumstantial evidence suggesting that viruses such as cause canine distemper or rinderpest have been extremely important factors in determining the distribution and abundance of wild terrestrial vertebrates in Africa or North America. Similarly, rabies epizootics in Europe have been characterized by diminished red fox populations, a consequence not only of the virus's pathogenicity but also of intensified hunting which attempts to lessen the risk of human infection. Poxviruses also seem to have had drastic effects on birds in the Hawaiian islands (Warner, 1968), where they were probably only recently introduced, but not in wild birds of North America, e.g. bobwhite quail (*Colinus virginianus*), where only 1% lethality was estimated (Davidson *et al.*, 1980). Perhaps the North American experience indicates long-established endemicity of avian poxviruses.

D'Herelle (1926) was strongly of the opinion that the progress of epidemic bacterial diseases in humans was determined at least in part by virus attack, but the importance is now thought to be trivial in humans although it might be greater in concentrations of farm animals. Even though *in vitro* tests have shown that viruses lessen the efficiency of nitrogen fixation, there are no compelling data implying that *in vivo* viruses significantly affect either the natural populations of nitrogen-fixing rhizobia in plants/soil or modify the progress of animal diseases (e.g. typhoid, cholera) caused by enterobacteria. Higher-plant population dynamics have normally been explained in terms of biotic and climatic factors interacting with population genetics. Although secondary effects of myxomatosis on rabbit populations and hence grazing of natural vegetation have been recognized, the possible importance of viruses in wild plants studied by ecologists has been neglected, largely because means of detection and characterization were lacking. Col-

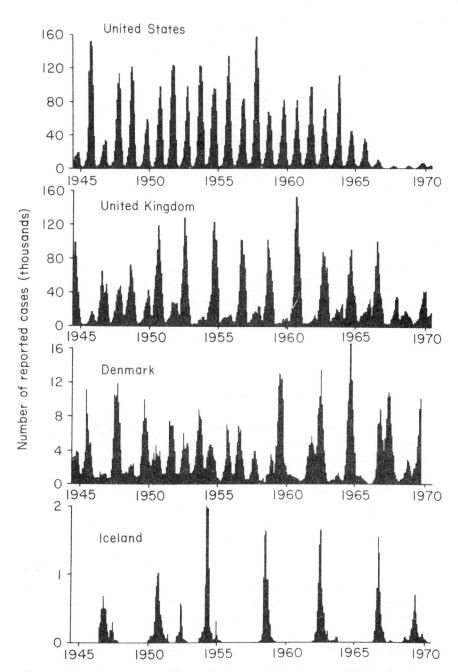

Figure 6.7 Reported cases of measles per month, 1945 to 1970, for four countries arranged in decreasing order of population size. Note the cyclicity, the
dramatic diminishment in amplitude for the United States after 1964 (because
of vaccination), and the marked non-endemicity in Iceland. (Reproduced with
permission from Cliff *et al.*, 1981.)

laborations between field ecologists and virologists have been lament-
ably few. However, the experience of plant breeders seeking wild
sources of virus resistances indicates that viruses probably do direct
changes in populations of hosts. In regions where specific viruses have
been endemic for centuries, the local crop cultivars show much less
disease when infected than more modern cultivars selected elsewhere.
Interestingly, diminished virulence of fungal plant pathogens thereby
influencing tree survival has been plausibly attributed to mycoviruses
spreading within natural populations of the fungi (Brasier, 1990).

Analysing the compartment model (Figure 6.6) in terms of equations
accommodating additional biological facts, Anderson and May (1979)
listed many of the criteria determining the ability of viruses to regulate
populations. Generally, if a virus regulates the population, the case
mortality, α, must be high relative to the net growth rate $(B-d)$ of the
virus free component. Any degree of regulation is lessened by lasting
immunity (γ must be small) and high rates of recovery after infection (R
must be large, as when infections are of short duration). Viruses
lessening rates of reproduction in infected components of the population
have an enhanced prospect of suppressing population growth. When
viruses are transmitted vertically they necessarily contribute to the
infected rather than the infectible parts of a community, thereby
lessening the equilibrium population within which virus is maintained,
even while it is a regulating factor.

A variety of other predictions is possible from the dynamic equation
in which the virus always becomes endemic in the sense that the
population, including potential hosts, tends to grow exponentially to a
level where the total population ($N = X + Y + Z$) exceeds a maintenance threshold determined by the various death and recovery rates
modified by the transmission rate. When this occurs, the fraction of the
population infected approximates to the net growth rate $(B-d)$ divided
by the virus-attributable death rate if the virus controls the population.
Alternatively, if the population continues to grow, the fraction depends
on the net growth rate minus the total population's growth rate (which
is necessarily less than the exponential growth rate of the virus-free
population) modified as before by the virus-attributable death rate. For
a fixed net growth rate, the degree of depression of the population's
equilibrium is modified by parasite pathogenicity. Increasing rates of
parasite-induced mortality result in a greater depression in the population until the rate of loss of infectors begins to have a detrimental
effect on the efficiency of disease transmission. When the virulence is
very high, infected hosts die before virus transmission occurs and the
virus is unable to be maintained within the population. Thus, the most
highly pathogenic (virulent) viruses killing their hosts at rates greater
than the virus's turn-round time cause their own extinction, but not

that of their potential hosts. This phenomenon has implications for strategies of biological control. Highly virulent (lethal) viruses are likely to have only transient effects on target populations; in the long run lesser virulence may be more effective in suppressing growth.

This is a special case of the ecologically accepted assumption that disease-free populations grow to an environmental-carrying capacity K, where K is the difference between *per capita* birth and death rates modified by density-dependent constraints (such as nutriment supply) on population growth. When the net growth rate of the population is very much smaller than the case mortality rate, α, introduction of the virus results in a classical epidemic in which the fraction of the total population infected peaks as the number of infectibles in the population diminish to a level that limits their prospect for contracting the virus. In terms of time, infective individuals are merely regenerating replacements for themselves but, as the proportion of infected individuals increases still further, it becomes less and less likely that they will contact an infectible and the case incidence drops at an increasing rate. After an interval determined by the fecundity of the survivors, the infectible part of the population increases and is substantially free from infection. It may take several cycles of virus regeneration before the proportion of infectibles increases to a level great enough to precipitate the next epidemic sequence which once again results in the decline in total population. Because the rate at which new infectibles appear is directly tied to the magnitude of the total population, the latter is closely correlated with prospects for maintenance. However, if the net growth rate of the population is very much less than the death rate (or the recovery rate tied with lasting immunity in vertebrates) it is likely that the incidence of virus infections after one epidemic event is so small as to allow accidental elimination of virus from the system.

Seasonal variation in transmission efficiency superimposed on the stochastic influences (chance) may enhance the prospects of liberating populations from viruses which they can maintain. Although the reasons are unknown, the seasonality of measles virus in human populations is well defined, so that only 1% of the total number of recorded infections occur in the period August to December. This has implications both for virus maintenance on the one hand and eradication on the other. Nathanson et al. (1978) used an essentially similar series of deterministic equations to those used by Anderson and May, to construct the seasonal waves in the percentage of New York's urban populations that were infectible with measles. In doing that they took into account the copious data available on the virus incubation period (before infecteds become infective, c. 12 days), average daily birth rates, death rates, case rates, etc. and assumed permanent immunity. Their computer accountancy method emphasized the seasonality of transmission, which they found to peak in December–January, some months before the case incidence peaked in the following summer (Figure 6.8).

More interestingly, by simulating case incidence of measles in populations containing differing proportions of infectibles, they found that when the level of infectibles exceeded 6%, the virus was likely to be maintained. It is noteworthy that the seasonal trough in transmission efficiency has a cumulative effect which depends upon the turn-round time of the virus. Yorke *et al.* (1979) commented that if the transmissibility of measles was diminished over a 60-day period, the effect would be manifest over five (12-day) regeneration periods from infected to infective. If case incidence diminished by a factor of 0.7 per generation over the 2-month period of lowest transmissibility, the resulting decrease in measles would be 0.75. For a virus such as influenza A, with a very short regeneration time (*c.* 3 days) and seasonal variation in prevalence, the effect would be manifest as a 0.7^{20} (*c.* 100-fold) decrease in incidence. For this reason they speculate that influenza virus A might be unable to persist in any human population without the advantages conferred by antigenic variation (drift/shift, see pp. 114–6).

6.7 THE INFLUENCE OF IMMUNITIES ON VIRUS MAINTENANCE IN POPULATIONS

In natural populations, immune components are either the direct consequence of natural infection or genetic change (Fraser, 1990). However, in humans and domesticated animals/plants, two additional immune components may be included in the populations (Figure 6.9). When plants are virus-infected, this sometimes protects against later infection by more severely damaging isolates of the same virus. This phenomenon has been applied to a few horticultural crops including tomatoes (Rast, 1972), cucurbits (Wang *et al.*, 1991; Walkey *et al.*, 1992), cocoa (Hughes and Ollenu, 1994), citrus (Bar-Joseph, 1978) and passion fruit (Simmonds, 1959), but the epidemiological consequences have not been the subject of detailed mathematical treatment.

Cross-protection in plants has a counterpart in bacterial and vertebrate virology and is sometimes attributed to dominant-negative mutation. Dominant-negative mutants have been recognized in a wide range of viruses; durable persistent infections seem particularly fruitful sources perhaps explaining why long-lived perennials have been sources of natural cross-protecting inocula (e.g. *Theobroma cacao* for swollen shoot badnavirus). The protective tomato mosaic tobamovirus inocula which are used commercially were manufactured (and selected) after nitrous acid mutation. As a consequence, the genetic difference between this mutant and the wild viruses against which it protects is

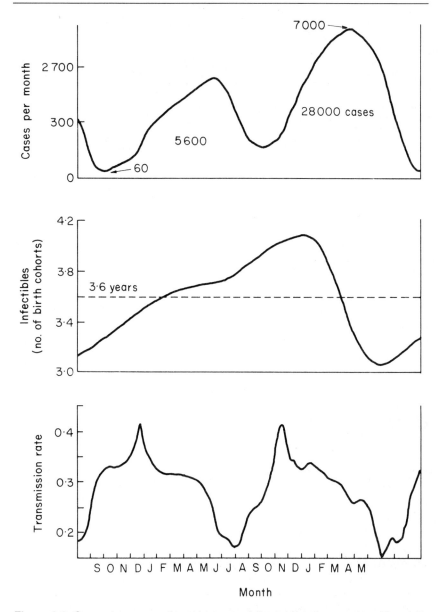

Figure 6.8 Computer accounting of seasonal fluctuation in measles. The daily number of cases (top panel) has been used to generate the curves showing seasonal variation in number of infectibles and in transmission rate. The high and low years are based on reported cases for New York City, and it is assumed that 16 000 infectibles enter the population annually, equivalent to a birth rate of 20 per 1000 in a population of 840 000. (Reproduced with permission from Yorke *et al.*, 1979.)

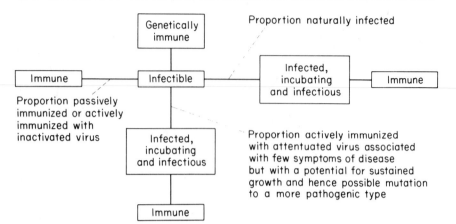

Figure 6.9 Diagram showing the natural and artificially induced immune components of a population which collectively modify the rate and extent of epidemics.

known to be a modest base change. However, the genetic distance between natural (cross) protective viruses and their wild targets is unknown. The term dominant-negative derives from the phenotype and reflects the relative abundance of the effective genotype in the population and its ability to coexist stably with other genotypes in the population. This underlines the fact that cross-protection is not an absolute exclusion – merely tolerance based on some poorly understood process of interference. Nevertheless, the phenomenon tends to support the view that disease severity reflects virus concentration. Field exploitation of cross-protection requires care. Use of mild isolates near genetically resistant plants facilitated virus evolution towards enhanced virulence (Pelham *et al.*, 1970; Pelham, 1972) and lack of durable protection, partly attributable to uneven distribution particularly in trees (Bar-Joseph, 1978) is another negative feature.

Mild strain protection in plants has its counterpart in vertebrates as live vaccination (with dominant-negative mutants – often attenuated during numerous passages in unusual hosts, at abnormal temperatures or, with more uncertainty, in genetically resistant hosts). Antiviral therapy (Whitaker-Dowling *et al.*, 1991) involving expression of genetically modified dominant-negative mutants (sometimes called intracellular immunization) is only now being assessed in vertebrates but in plants this approach (termed pathogen-derived resistance) (*sensu* Sanford and Johnston, 1985) has been used for more than a decade.

The expression in transgenic plants of viral capsid proteins was associated with tolerance to subsequent challenge by related viruses (including those delivered via vectors: van der Vlugt and Goldbach,

1993; Cooper *et al.*, 1994). It is now known that these products are not essential; virus-derived incomplete structural or non-structural genes sequences, coding sequences, non-coding sequences or untranslatable sequences in sense or antisense orientations have all been reported to cause tolerance. In a very few instances the protection which has been observed when a viral gene has become a plant gene has extended somewhat beyond that of the source virus (Wilson, 1993) and the phenomenon has been studied in relation to a diverse array of economically important virus taxa: tospoviruses, ilarviruses, carlaviruses, cucumoviruses, luteoviruses, nepoviruses, potyviruses, potexviruses, tenuiviruses, tobamoviruses, tombusviruses and tobraviruses (Fitchen and Beachy, 1993). Although the mechanism of action is unknown, technology push is likely to cause extension of such transgenic crops, initially on a trial basis, into geographic ranges where production is presently constrained by a lack of tolerance to indigenous viruses. A contrasting situation can be recognized when the mutant suppresses growth of a related wild-type virus and becomes the dominant component in the population (hence genotypic dominance). Imbalance in amounts of structural units brought about by prematurely terminated translation may be sufficient to cause the dominant phenotype to emerge; their action is suspected to be competition or blocking of a scarce resource. Thus, defective interfering RNAs and satellites, being smaller than the viruses which aid their replication, more successfully compete for polymerase complexes, thereby downgrading replication of the complete genomes. As outlined, there are new plant genotypes in the periagricultural environment and the current range of virus-derived tolerance genes is soon likely to be augmented by transgenic genes from several other sources – notably plants or mammals (albeit after many cycles of amplification in bacteria and their viruses). Two broad classes of approach in the latter category are antibodies (planti-bodies) targeted on viral polymerases, movement proteins or other critical low abundance products (capsid proteins seem an unpromising target; Taviadoraki *et al.*, 1993) and sequences which are suspected to play a role in 'scavenging' of dsRNAs (e.g. 2'-5' oligoadenylate synthetase; Truve *et al.*, 1993). In addition, it is anticipated that two classes of plant-derived sequences will become available for use as transgenic genes: hypersensitivity (quasi-immunity) genes (e.g. the N gene which is effective against tobacco mosaic tobamovirus; Fraser, 1990) and, more speculatively, natural non-host genes. Traditionally, breeding of agricultural and horticultural crops for virus tolerance/resistance/immunity exploited natural biodiversity in wild relatives. A variety of simply inherited major genes was brought into commerce from these sources and the scale of searches for the desired character gives a crude measure of their abundance in nature. There is a mass of data on viruses of potato and the prevalence of simply inherited major genes for virus immunity/tolerance seems low. It is difficult to guess the importance of this experience outwith the site of collection – the Andean regions of South America. However, a survey of horticultural crops revealed no known sources of resistance to 25 viruses affecting them and, in these plants the introduction of transgenic resistance traits might be anticipated to introduce a substantially new dimension into the dynamics of parasite evolution. When the transgenic genes are virus-derived an additional novelty can be envisaged because their presence throughout a crop removes the element of chance (reflecting vector preference) which otherwise constrains opportunities for multiple infection and its evolutionary consequences.

The practice of deliberately infecting vertebrates with viruses of low pathogenicity has been routine since the eighteenth century when Edward Jenner publicized the protection (to individuals) afforded against variola (causing smallpox) by a distinct poxvirus then associated with cowpox disease. In human and veterinary medicine, protection of individuals and communities is achieved in several ways. One method resembles that used to protect plants: infectious (attenuated) viruses selected for their ability to elicit only mild symptoms of disease are used to confer durable immunity. Another method uses inactivated viruses which cannot grow in cells of the inoculated subjects and consequently elicit only transient immunities. Table 6.1 lists the con-

Table 6.1 Conditions of viruses in some vaccines of clinical or veterinary importance

Disease	Attenuated	Inactivated
Smallpox	+	−
Mumps	−	+
Measles	+	+
Rabies	+	+
Yellow fever	+	−
Rinderpest	+	+
Newcastle disease	+	+
Canine distemper	+	+
Foot-and-mouth	−	+
Poliomyelitis	+ (Sabin)	+ (Salk)
Rotaviral diarrhoea	+*	−
Cytomegaloherpes	+*	−
Hepatitis A	+	+*
Influenza	+	+
Dengue	+*	−
Japanese encephalitis	+*	+

*Under development.

ditions of viruses in some of the vaccines of clinical or veterinary importance.

In vertebrates, immunity may additionally be acquired passively consequent upon the administration of antibodies made in different individuals, as when antibodies are transmitted from mothers to off-spring *in utero* via the placenta and/or via the milk (colostrum). Alternatively, transient passive immunity can be conferred when serum globulins from an immune individual are injected into another to protect, or exceptionally to abort, infections. The extent to which these various immunities affect the dynamics of epidemics in vertebrate populations has been simulated on numerous occasions to determine the cost-effectiveness (Frerichs and Prawda, 1975) or in other ways to optimize vaccination programmes. For example, the studies of Yorke *et al.* (1979) emphasized the potential for measles eradication from large urban populations by using immunization to lessen the proportions of infectibles. Finding that the magnitude of the infectible population at the time of seasonally low transmission efficiency is the determining factor in virus maintenance in human populations, they recommended vigorous intervention aimed at early virus (disease) diagnosis and vaccination of infectibles in the vicinity. Because the target population in children is about 4 years of age, it is more or less readily available in pre-school playgroups, etc. Undoubtedly other viruses with seasonal cycles of prevalence and an obligate requirement for new hosts to replace those lost due to acquired immunity will be similarly amenable to contrived rather than chance eradication (cf. Anderson and May, 1982a,b).

6.8 SOME PROBLEMS IN MODELLING VIRUS—MULTICELLULAR HOST SYSTEMS

It is in some ways surprising that the non-specific models, in which species growth and interactions and the factors responsible for stabilizing the equilibria or limiting cycles are described in terms of simultaneous quadratic equations rather than in strictly biological terms, accord so well with biological experience. In the context of virus biology it is particularly noteworthy because numerical assessment of infectible cell populations which is possible with unicellular organisms becomes much more complicated when the subjects are multicellular. Not all cells in a multicellular organism are equally exposed, infectible or sensitive to viruses in their vicinity. Furthermore, tissues in which specific viruses can grow are probably by no means uniform; the cells that form them may well be infectible during only part of their development. Thus, the numbers of infectible organisms is only a rough guide to the number of infectible sites. Furthermore, the volume occupied by the infectibles, and hence their exposure to viruses, is liable to vary in response to a variety of factors such as nutriment supply. Plants are particularly variable in this regard but it is generally true that the potential for exponential growth (implicit in the mathematical treatments of population dynamics) is by no means closely tied to numbers of multicellular individuals.

6.9 VIRUS DYNAMICS IN UNICELLULAR POPULATIONS

With bacteria, theory and experiment have gone hand-in-hand in the evaluation of models based on specific assumptions regarding habitat, resource use, population growth and the conditions facilitating stable states of coexistence with their viruses, bacteriophages (phages). Bacteriophage interactions have given rise to a profound literature and a plethora of specialized terms. Consequently it is necessary to digress so as adequately to set the scene.

Because of commercial sensitivities about intellectual property and processes, details about the impact of viruses on modern industrial reactions are not readily available. However, more traditional industrial microbiological processes, such as the manufacture of cheese, antibiotics or butyl alcohol, periodically suffer irritating inconvenience when viruses are accidentally introduced into the bacterial populations on which the processes depend. The advent of genetically manipulated fermentation for the production of bovine somatotropin, insulin, blood clotting factors, etc. will probably come into conflict with viruses. Reliance on a single starter culture (genotype) makes the system vulnerable and calls for expensive design, operation and monitoring procedures. The economic impact of bacterial viruses has rarely been great enough to justify much research, and there is a paucity of information concerning phages in 'wild' bacterial populations, although their occurrence in sewage, soil, water, and in and on plants and animals is documented. Now that there are prospects of genetically engineered bacteria on plant surfaces to minimize frost damage or potentially in the roots to enhance nitrate utilization, a variety of tests has been carried out to confirm suspicions that gene flow is facilitated by viruses in the farmer's field as well as in the laboratory culture. In the pursuit of basic (specifically genetic) knowledge, phages have been objects of intensive investigation because of their 'small' genomes and particularly because of the rapid regeneration and easy maintenance of their hosts. Although phage particles have been recognized in Gram-positive cocci (e.g. *Staphylococcus, Streptococcus*) as well as in bacilli such as *Clostridium, Bacillus* spp., most is known about phages that infect Gram-negative bacilli, notably Enterobacteria and pseudomonads (Reanney and Ackermann, 1982). Phages include viruses containing ssDNA (e.g. Inoviridae), dsRNA (Cystoviridae) and ssRNA (Levivi-ridae). The diversity in shape and apparent sophistication is consider-able, and many phages are designated by numbers or Greek symbols. Among the most sophisticated are the T even group of tailed phages containing dsRNA (Myoviridae) which have been notable models in systems which have given insight into virus–host dynamics on the one hand and options for virus maintenance on the other. Even though much of the discussion that follows relates to Myoviridae or phages such as λ (Siphoviridae), it is important to remember that they are by no means typical. Indeed, by possessing a specialized nucleic acid-injection system and a contractile tail, Myoviridae seem to be highly evolved. With few exceptions, bacterial viruses lacking tails must attach to and infect via projections (pili) on the surfaces of potential hosts (Figure 6.10).

Interestingly, these bacteria with pili are in anthropomorphic terms male; the pili are required for bacterial conjugation and are a mani-festation of a fertility factor (F) which is itself infectious but only by direct cellular contact. Even though the molecular biology of bacterial viruses has been well described (Stent, 1963; Hayes, 1964) it is appro-priate to restate the two natural states in which phages may occur in one narrowly defined host phenotype:

1. Lethal (virulent) states in which phages infect bacteria, replicate and cause host cell lysis.
2. Temperature states in which phages additionally multiply indefinitely within bacterial cell lines. The non-infecting intracellular state of the phage is called prophage and in this covert form phages have been found to be almost inescapable contaminants of all bacterial po-pulations that have been studied with the appropriate techniques.

When a temperate phage infects a cell, the host undergoes a per-manent hereditary change and gives rise to a lysogenic cell lineage with two other properties:

• A standard proportion per bacterial generation bursts to liberate descendants of the original phage.
• The bacterial cells do not normally succumb to infection resulting in lysis with phages of the genotype that earlier infected their progenitor.

Thus, at least some bacterial viruses are intimately maintained with-in bacterial populations to which they confer protection against superinfection.

In their natural environments (e.g. intestinal tract of vertebrates, plant-root interfaces with soil water) bacteria are bathed in solutions which are constantly renewed in the sense that there is continuous

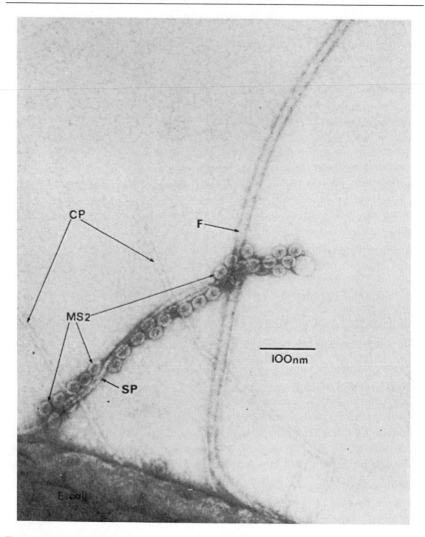

Figure 6.10 Transmission electron micrograph of *Escherichia coli* showing part of a sex pilus (SP), to which spherical levivirus particles (phage MS2) are attached. Virus-like particles are not associated with a flagellum (F) or common pili (CP). Print provided by Dr A. Lawn, Lister Institute of Preventative Medicine as used in Hawker and Linton (1979).

replenishment of nutriment and continual removal of waste products of metabolism as well as 'spare' phages. The system is therefore different from that in which bacteria are reared routinely in laboratories: on batches of media containing finite quantities of nutriment controlling the final yield of cells. In continuous-culture systems (chemostats), bacteria are effectively starved of nutriment; the bacterial generation time is determined by the rate of nutriment flow, with each new drop of food being used instantaneously. From experimentally measurable properties such as rate of resource conversion, flow rate, numbers of phages per bacterium and hence numbers and time delay in producing new phage populations, Levin et al. (1977) developed a model of predator–prey interaction analogous to those mentioned above but differing somewhat in mathematical structure. It took account of prey (bacterium) growth in relation to the availability of resources, and the expectation of stable states of coexistence between Escherichia coli (B strain) and a lethal phage (T₂) is particularly interesting and probably relevant to a variety of planktonic species. When the theory was put to the test, technical problems, including a 5% variation in flow rate (resource input), contributed to wild oscillations in bacterial and phage populations and these led to the elimination of the bacterial population. Nevertheless, there was a good agreement of real with theoretically predicted equilibrium levels when a single potentially limiting resource, glucose, supported a single bacterial population (with a very low rate of mutation to phage immunity) which was preyed upon by a single phage population. Analogous problems thought to be due to patchyness in their system prevented B.R. Levin and his colleagues in Emory University, Atlanta, Georgia (personal communication) from completely explaining the population processes which underlie therapies in which antibiotic treatments are coupled to phage therapy when attempting to control sepsis in vivo. Nevertheless, Levin and his co-workers were able to support experimentally the classic experiences of Smith and Huggins (1982, 1983), in mice and other animals (Smith et al., 1987).

Interestingly, Levin et al. (1977) extended their investigation to one in which two primary consumers (bacterial populations) interacted with a phage that infected only one of them, a situation that would be expected to evolve as a consequence of natural genetic processes and one that had indeed been observed by Levin et al. and others. To validate their theoretical predictions, the E. coli (B) and T₂ phage were grown with a T₂-immune clone of E. coli (K12, strain S30). The two bacterial populations differed in another respect; E. coli K12/S30 was at a competitive disadvantage relative to E. coli B; the introduction of E. coli B into an equilibrium phage-free chemostat culture of K12/S30 rapidly resulted in an increased frequency of E. coli B. Starting with equal numbers of the two bacterial clones in the presence of phage there was a rapid decline in the concentration of E. coli B and in the T₂ phage, while the immune clone increased in abundance until it approached the concentration that would have occurred had it been alone in the chemostat. However, in time, the E. coli B population recovered and continued to increase, with the phage population keeping in step. This qualitative demonstration of short-duration (400-hour) coexistence was in accord with these models, i.e. when the second-order consumer (of glucose in the carbon energy source) is at a disadvantage with respect to resource use but immune from phage, and if the second-order species can survive on the amount of food left over when the two first-order consumers (*E. coli* B and phage T_2) are in equilibrium, then a stable state is possible with all three present.

In reality, additional evolution has been observed in phage-bacteria interactions. When phage-immune bacteria are numerous, the mutants in the phage population are liable to be selected, and phage with extended host ranges create environments selecting for bacteria with new orders of 'immunity'. Interestingly, one study revealed the apparently stable coexistence of a phage (T7) with an infectible *E. coli* B population, with broader host range than its progenitor. Thus, although spatial and temporal heterogeneity undoubtedly contribute to the stability of bacteria-phage populations, coexistence is possible even in the uniform habitats implicit in many theories of predator–prey interactions. Indeed the introduction of a virus into a quasi-natural situation of species competition can force coexistence when it might not otherwise occur. In this respect, experience gained from experiments with bacteria and phages is in accordance with classical predator–prey studies (Gause and Witt, 1935). Furthermore, there is evidence that viruses may be maintained within populations, even while debilitating or killing the host because there is complementary benefit accruing from the elimination of some of the host's competitors. One way this can happen is when a population including infectibles is capable of acting as a carrier/reservoir of a virus that is more damaging in an alternative host. Although the foregoing has given some qualitative insight into the coexistence of bacteria with their viruses, the mechanisms and the implications remain to be assessed. It is unreasonable to assume that the only mutants produced during a period of host–virus coexistence will be those endowing second-order hosts with immunity from infection. This selective advantage in the closed model may be no greater than one deriving from a completely different enhancement of fitness to survive (e.g. efficiency of resource utilization) which evolves in a competitor within the community. In a closed system it is likely that the phage-infectible population will be eliminated before the notionally more adapted mutant establishes its equilibrium within the finite total population and the virus will die out as a result.

6.10 THE OPTIMUM MAINTENANCE STRATEGY: SPECULATION ABOUT THE COSTS AND BENEFITS OF CO-EVOLUTION

In the natural environment, many viruses are maintained not only by association within hosts which die out as a consequence but also by becoming intimately associated with the genomes of hosts that continue living. The ultimate maintenance strategy seems to be one of integration into the host's chromosomal core, an option restricted to viruses having DNA genomes (herpesviruses, papovaviruses, adenoviruses, parvoviruses; Croen and Straus, 1991) or those such as retroviruses having an ability to convert their genetic information between two states: RNA (in virus particles) and DNA when integrated for perpetuation within their hosts. In bacteria, virus integration (as prophage in lysogenic populations) has been studied in most depth and profound consideration has been given to the consequences. Superficially, integrated prophage is a nuisance to its host, which must divert resources towards the synthesis of the foreign nucleic acid. Since this would tend to be disadvantageous and thereby liable to result in the demise of hosts (through competition) as well as the virus, compensations have been sought. Advantages to the hosts have been recognized both at the cellular and the population levels. As mentioned earlier, lysogenic cultures tend to be immune from superinfection. More interestingly, lysogeny has been found to: (i) alter the surface properties of *Salmonellae*; (ii) render harmless isolates of *Corynebacterium diphtheria* more damaging to their human hosts by stimulation of toxin production (Betley *et al.*, 1992); and (iii) regulate the permeability of coliforms, thereby facilitating glucose uptake. These phenomena, collectively termed lysogenic conversion, have been shown to confer improved survival potential in chemostat cultures. In one series of experiments Edlin (1978) showed that lysogenic populations of *E. coli* containing prophage λ have a reproductive advantage in aerobic but not anaerobic conditions as illustrated in Figure 6.11.

As yet, there are few data implying that viruses enhance the survival prospects for the genes of their multicellular hosts. A notable exception is a report by Gibbs (1980) who observed that when an Australian wild plant (*Kennedya rubicunda*) was infected with a tymovirus, rabbits ate less of it than virus-free material of the same species. Another dimension is provided by attempts to explain advantages conferred by sexual reproduction. Theoreticians have proposed that parasites represent the major selective force favouring the maintenance of genetic variation and sexual reproduction in plants as well as animal populations (Hamilton, 1990; Lively *et al.*, 1990). Recent data (Kelley, 1994) support the view that barley yellow dwarf luteovirus infection generates advantages for sexually reproducing genotypes in sweet vernal grass (*Anthoxanthum odoratum*) communities. For infection to be important in this context, a virus (or perhaps viruses more generally) should lessen the performance of individuals (whether or not symptoms are overt) and should also be a prevalent, although not necessarily, uniform force. Many studies show that wild species are frequently infected by diverse viruses and measurably suffer when infected (e.g.

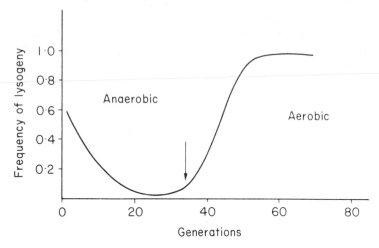

Figure 6.11 Reproductive ability under anaerobic and aerobic conditions. *E. coli* strain AB and a lysogenic population (ABλ) grown in a glucose-limited chemostat under anaerobic conditions. Then (at arrow) culture switched to aerobic growth: the lysogenic (phage–bacterium) population gradually replaced the non-lysogenic cell lineage achieving an equilibrium after some 20–40 generations.

Anthoxanthum and brome mosaic bromovirus: Kelley, 1993; *Betula* and cherry leaf roll nepovirus: Cooper *et al.*, 1984; *Eupatorium* and tobacco leafcurl geminivirus: Yahara and Oyama, 1993).

Fluctuating microclimate can induce chaotic flux in parasite populations (Hassel *et al.*, 1991) but conventional wisdom (Begon and Harper, 1989) anticipates the evolution of parasitism tending towards commensalism. However, other pressures were considered by Ewald (1983) to be capable of driving evolution towards severe pathogenicity. On this basis, vector-transmitted viruses tend to evolve towards benign parasitism or commensalism in vectors and severe pathogenicity in alternative hosts (plant or vertebrate). In parallel, vertically transmitted parasites should evolve towards benign parasitism or mutualism. The transmissibility of the parasite (virus) is a crucial testable component. An examination of current knowledge concerning viruses which infect plants is in line with these predictions. By contrast, selective advantage from associations between host populations and virus has been frequently reported in the bacteriological literature. Phages can carry bacterial genes from one host to another, permanently altering the genetic constitution of the recipient even though the phage need not become lysogenic as a consequence. This potential to incorporate bacterial genes in phage (transduction) is undoubtedly beneficial in facilitating natural survival and the phenomenon may not be restricted to bacteria. Experimentally, other viruses such as cauliflower mosaic caulimovirus are routinely used as parts of chimeral vectors of genes between angiosperms and it is probably justified to consider them as 'natural genetic engineers'. Whether or not the vector potential of DNA viruses is general, it is interesting to note the blurring effect it has on classical genetic and species concepts. Within the bacterial cell–DNA phage system, genes can be interchanged, rendering a particular piece of DNA 'bacterial' in one environment and 'phage' in another. Such considerations at the interface between virology, evolution and genetics were fascinatingly and convincingly expounded by Campbell (1961, 1981) and Reanney (1978).

Although some DNA phages become integrated within the genomes of their hosts, other phages containing DNA are not known to be so intimately associated with the genetic core and run a risk of being lost from the germ line during division of cells in which they occur. In this respect non-integrated phages resemble the diverse array of viruses containing RNA (except retroviruses) and other viruses containing DNA (e.g. caulimoviruses) which have not been detected in the chromosomal DNA of their multicellular hosts. Since the non-integrated viruses have been associated with numerous effects (mostly harmful) they can be likened to a floating pool of genetic variability which is liable to be expressed in the phenotype of their hosts. Interestingly, the virus DNA that floats in this way forms part of a spectrum of nucleic acid species which were separately discovered as normal cell constituents to which geneticists gave special names such as plasmids, transposons, etc. Although both transposons and plasmids contribute information that may be expressed in a cell's phenotype, they are functionally distinguishable from one another: the former seem able to migrate from one part of a bacterium's genetic core to another and the latter additionally possess an ability to transfer genetic information from the 'chromosome' of one individual to others, being transmissible vertically and/or horizontally. The conjunctive plasmid (Hfr) conferring 'maleness' to bacteria is in this category. In common with viruses that are not integrated, each of these floating nucleic acids replicates to maintain itself so that the occurrence of a plasmid in a new cell is not normally accompanied by loss from the progenitor. With the recognition that at least some of the floating nucleic acid species were capable of replication at rates greater than those of the chromosomal core nucleic acids of their hosts (immediately before intracellular transfer), it became apparent that, even though theories about maintenance of viruses originated against a different conceptual background, the discussions converged. Plausibly, plasmids, etc. are highly adapted invaders in which rates of autonomous replication are controlled so as to maximize benefit to both the cells and themselves. Because maintenance of the floating nucleic acid is a metabolic burden tending to lessen the efficiency of its host, compensating benefits have been anticipated and there is a continuing search to identify them. Additionally, because of the possible relevance of this information to the management of cancers that have been associated with viruses, work is being directed towards the identification of mechanisms of control. The massive literature indicates that control may be exerted either by the host's genetic core or by the floating nucleic acids themselves. Temperate prophages such as λ are known to encode repressors that specifically turn off transcription from the virus genome so that the potential for excessive autonomy is not expressed. Furthermore, some conjunctive plasmids are associated with repression of pilus formation, thereby lessening the prospect of their cells being superinfected with viruses (notably phages containing RNA) that exclusively infect via pili.

Thus, within a cell, gene flow is possible between the chromosomal core and naturally occurring nucleic acids floating in the cytoplasm. Furthermore, conjunctive plasmids and nucleic acid species attributable to viruses facilitate natural genetic engineering by having the capacity to carry genes from one cell to others in the population. There are, as yet, few data on the extent to which gene flow between hosts having different core genotypes is possible in nature but at least with the most studied laboratory populations of bacteria, the effect of this tendency to redistribute seems small. The genetic maps (chromosomal core DNA) in Gram-negative bacilli such as *Escherichia* and *Pseudomonas* are distinctly different from one another and from the more obviously different Gram-positive bacilli. Presumably there are many genes which are essential for survival and any tendency for these to be relocated or lost is more harmful than the benefits presumed to accrue from floating genes that occur in tandem. If one accepts that the floating genes are dispensable and that a cell can exist using only its chromosomal core, it is possible that floating nucleic acid species have a speculative role in a cell's strategy for survival. Features that they confer may be appropriate only in situations that recur infrequently as when they code for properties giving their hosts antibiotic resistances. However, when the host's changing environment stabilizes, there is the chance that the experimental attributes will become integrated into the chromosomal core. Conceivably, genes may revert to their floating role following additional changes in the host's environment. In time, a population may lose particular floating genes and/or may acquire new ones from those circulating in the community.

Whereas such a scenario accords with experience gained concerning the movement of DNA between bacterial populations, there are few compelling data supporting this possibility in multicellular organisms. Multicellular organisms necessarily exhibit a greater potential for differentiation of cells that are clonal descendants of the sperm/egg fusion product. This gives a variety of phenotypes that is attributable only in part to the proteins coded for by the dsDNA of chromosomal cores. Recognizing that cells are controlled by DNA through the medium of RNA species (which are translated into proteins), from time to time, transcription error presumably generates redundant RNA species. In some instances these RNA species may be harmful (like viroids) or inconsequential, but where improvement in the fitness of the cells results, an increasing diversion of RNA metabolism along the new pathway may be possible. Nahmias and Reanney (1977) have speculated along these lines, envisaging an origin for some groups of RNA viruses from transcribed nuclear RNA species. Whether this hypothesis is correct or not, the traditional concept of polynucleotides as fixed entities seems untenable. Nucleic acids are in a constant state of flux and viral genomes probably form only part of a continuum which may be harmful or beneficial depending upon the cellular environment in which they occur.

Conclusion

Speculation about the co-evolution of viruses with their hosts was justified because of relevance to mechanisms of virus maintenance. It is inconceivable that the diversity and sophistication of those viruses about which much is known have had only brief evolutionary histories. Nucleic acid fits the description 'selfish' and, having the capacity to replicate for maintenance at rates somewhat faster than that of host chromosomal DNA, it is not hard to anticipate that environmental circumstances will sometimes facilitate greatly increased over-production of nucleic acid species which thereby acquire some of the attributes of a virus. In some instances this over-production may result in death of host cells. Provided that suitable nearby cells are replaced at a sufficient rate one imagines that horizontal spread by contagion alone will be adequate for the survival of the nascent virus but, outside the cell, nucleic acids are vulnerable to extracellular enzymes and possession of a protective protein coat will be advantageous and likely to be a property selected for. Monophagy, deriving from a close co-evolution with one type of host, is likely to be the first option in the evolution of the nascent pathogen, but oligophagy would provide greater security and adaptation to the natural environment's patchyness. One way in which this goal may be achieved is by chance association with a vector. Viruses endowed with proteinaceous coats and achieving high concentrations in cells that they infect are probably the most likely to be casually associated with potential vectors when they feed, e.g. mechanical transmission of the virus of myxomatosis by mosquitoes. Similarly, the availability of virus particles in soil water renders them liable to analogous recovery by fungal zoospores. Reflecting the feeding preferences of the casual vectors, the viruses are liable to find themselves deposited in some tissues of multicellular organisms more than others and in these circumstances there will be a tendency for adaptation to specific tissues/organs. Alternatively, tissue tropism may reflect avoidance of humoral immune systems in vertebrates. However, in those instances in which tissue tropism enhances availability to vectors, there is likely to be an increased selection for parts of the virus population that are most efficiently acquired. Selection pressures probably encourage evolution to points where the viruses elicit the production of compounds that facilitate vector association (as with potyviruses of plants). With the increased mobility attributable to vector association, virus populations acquire enhanced prospects of accidental association with viruses having different evolutionary histories from their own. One consequence of resurfacing and genome resortment is increased vector or host range (as with luteoviruses of higher plants; Rochow, 1972), but, recognizing the diversity of multicomponent viruses mentioned earlier, other properties are also liable to change. Evolutionary trends towards enhanced virulence is a theoretical possibility. However, since excessive virulence is counterproductive (from the virus's viewpoint) a balance must be struck between the selective effects of viruses on their hosts and vectors and countervailing forces. Being the more fecund, it is probable that viruses force the pace of co-evolution with their hosts. Perhaps all viruses are beneficial in those hosts in which they are maintained. Unfortunately, these maintenance hosts do not attract casual interest and the generality of the assertion is unlikely to be demonstrated by biased surveys of diseased organisms. It is predictable that some hosts which are now identified as reservoirs may not yet have achieved the optimum mutualistic relationship with viruses that they carry. On the other hand, some hosts in which viruses cause debilitation may be evolutionary dead-ends. The most successful animals/plants in long-established natural communities now seem to be an appropriate group of targets for virological study. In the context of plants, these survivors may contain genetic determinants for non-infectibility by viruses and molecular knowledge of the putative locks that fit around specific viral keys may be exploitable in crops. Potentially, the viruses which have attained the optimal maintenance strategy (close association without integration in their host chromosomes) will have interesting stories to tell.

Recognizing the wide host range of most viruses and the opportunities which exist for recombination (Gal et al., 1992; Bacher et al., 1994; Gibbs and Cooper, 1995), real conflict of interest seems likely to emerge from attempts to reconcile conservation and management of wildlife for amenity/production with human welfare. One manifestation has been the reaction of elements in the lay community to the prospect of crops containing transgenic genes and the more general reservation about the ethical issues tenuously linked to novel foods and production systems. Genetically engineered viruses are now in the environment and people containing them are now a tiny part of the human population. Genetically engineered plants containing virus-derived sequences will not become commonplace if superior alternative approaches towards optimizing resource utilization are found. However, as judged at present, there is a real possibility that such novel genotypes will become an undetectable component of bulk commodities (cereal grains, soybean seeds) which are traded by the shipload. In these circumstances choice at point of sale will be unrealistic and, in a similar way, transgenic fish will escape the confines of their hatcheries and contribute to the natural biodiversity. The World's human population needs to be sustained while quality of life is maintained (ideally enhanced). Biotechnology, including the exploitation of viral resources, is justified – if it delivers what it promises. There may be no negative consequences from this scientific thrust but any that do occur are unlikely to be apparent in the short term.

Virology is a specialized branch of ecology that has made great strides – not always unambiguously good. The general advance of virological knowledge has allowed some of the most dangerous pathogens to be grown in sufficiently large quantities for weaponization yet the extent of information on the natural occurrence and potential of viruses is very limited. The central parts of South America, Australia, Antarctica, Africa and Asia are still virtually unexplored. It is predictable that the number of distinct viruses in a region is proportional to the number of potential hosts (although their recognition inevitably also reflects the number of virologists). The climatically inhospitable parts of the world such as the hinterland of Antarctica are unlikely to contain many. However, this is not likely to be true for the extensive and rich littoral flora and fauna of subpolar regions which have also been virtually ignored virologically. The greater average temperature and predictable climatic change of tropical forests facilitates biomass turnover and continuous, even distribution of resources which permits specialization (speciation) to numerous circumscribed niches. Primary forest zones seem likely to support the greatest diversity both of potential hosts and their associated viruses. The increased handling of wild-caught primates, coupled with urbanization, almost certainly facilitated the current lentivirus pandemic which is expected to involve 40 million people by the year 2000 with concomitant sociological burdens attributable to resurgence of tuberculosis and withdrawal of the agricultural workforce in developing countries. Terrestrial sources of viruses have long been recognized as potentially threatening. However, there have been repeated reminders in the past decade that the aquatic environments of the world may be no less fruitful: they are the most extensive, yet most neglected, virus ecosystems and Man is increasingly being forced to exploit them.

References

Abdalla, O.A., Desjardins, P.R. and Dodds, J.A. (1985) Survey of pepper viruses in California by the ELISA technique. *Phytopathology*, **75**, 1311.

A'Brook, J. (1964) The effect of planting date and spacing on the incidence of groundnut rosette disease and of the vector *Aphis craccivora* Koch at Mokwa, Northern Nigeria. *Ann. Appl. Biol.*, **54**, 199–208.

Ackerman, R., Bloedhorn, H., Kupper, B., Winkens, I. and Scheid, W. (1964) Uber die Verbreitung des Virus der lymphocitaren Choriomeningitis unter den Mausen in Westdeutschland. I. Untersuchungen uberwiegend an Husmausen (Mus musculus). *Zentbl. Bakt. ParasitKde* (Abt. 1)., **194**, 407–30.

Ackermann, H.W. (1992) Frequency of morphological phage descriptions. Arch. *Virol.*, **124**, 201–9.

Ackermann, H.W., Audurier, A., Berthiaume, L., Jones, L.A., Mayo, J.A. and Vidaver, A.K. (1978) Guidelines for bacteriophage characterization. *Adv. Virus Res.*, **23**, 1–24.

Adams, M.F. (1991) Transmission of viruses by fungi. *Ann. Appl. Biol.*, **109**, 561–72.

Ahlquist, P. and Pacha, R.F. (1991) Gene amplification and expression by RNA viruses and potential for further applications to foreign gene transfer. *Physiol. Plant.*, **79**, 163–7.

Allen, W.R. (1981) Dissemination of tobacco mosaic virus from soil to plant leaves under glasshouse conditions. *Can. J. Plant Pathol.*, **3**, 163–8.

Anderson, R.M. and May, R.M. (1979) Population biology of infectious diseases. Part I. *Nature*, **280**, 361–7.

Anderson, R.M. and May, R.M. (1982a) *Population Biology of Infectious Diseases: Report on the Dahlem Workshop on Population Biology of Infectious Disease Agents*, Springer-Verlag, Berlin, Heidelberg and New York.

Anderson, R.M. and May, R.M. (1982b) Directly transmitted infectious diseases: control by vaccination. *Science*, **215**, 1053–60.

Andrews, C., Pereira, H.G. and Wildy, P. (1978) *Viruses of Vertebrates*, 4th edn, Bailliere Tindall, London.

Andrews, G.L. and Sikorowski, P.P. (1973) Effects of cotton leaf surfaces on the nuclear polyhedrosis virus of *Heliothis zea* and *Heliothis virescens* (Lepidoptera: Noctuidae). *J. Invertebr. Pathol.*, **22**, 290–1.

Anon. (1979a) *Water Pollution Control Technology*, HMSO, London.

Anon. (1979b) *Human Viruses in Water, Waste Water and Soil*. Technical Report No. 639. World Health Organization, Geneva.

Armstrong, D., Forner, J.G., Rowe, W.P. and Parker, J.C. (1969) Meningitis due to lymphocystic choriomeningitis virus endemic in a hamster colony. *JAMA*, **209**, 265–7.

Atreya, P.L., Lopez-Moya, J.J., Chu, M. ,Atreya, C.D. and Pirone,T.P. (1990) Mutational analysis of the coat protein N-terminal amino acids involved in potyvirus transmission by aphids. *J. Gen. Virol.*, **76**, 265–70.

Ayres, M.D., Howard, S.C., Kuzio, J., Lopez-Ferber, M. and Possee, R.D. (1994) The complete DNA sequence of *Autographa californica* nuclear polyhedrosis virus. *Virology*, **202**, 586–605.

Bacher, E.R., Warkentin, D., Ramsdell, D. and Handcock, J.F. (1994) Selection versus recombination: what is maintaining identity in the 3' termini of blueberry leaf mottle nepovirus RNA1 and RNA2? *J. Gen. Virol.*, **75**, 2133–8.

Bahner, I., Lamb, J., Mayo, M.A. and Hay, R.T. (1990) Expression of the genome of potato leafroll virus: a readthrough of the coat protein termination codon in vivo. *J. Gen. Virol.*, **71**, 2251–6.

Bailey, L. (1981) *Honey Bee Pathology*, Academic Press, London.

Bald, J.G. (1960) Forms of tobacco mosaic virus. *Nature*, **188**, 645–7.

Baldwin, L.B., Davidson, J.M. and Gerber, J. (1976) *Virus Aspects of Applying Municipal Wastes to Land*, University of Florida, Gainesville.

Bar-Joseph, M. (1978) Cross protection incompleteness: a possible cause for natural spread of citrus tristeza after a prolonged lag period in Isreal. *Phytopathology*, **68**, 1110–11.

Barker, H. (1989) Specificity of the effect of sap-transmissible viruses in increasing the accumulation of luteoviruses in co-infections. *Ann. Appl. Biol.*, **115**, 71–8.

Baxby, D. and Paoletti, E. (1992) Potential prevalence of non-replicating vectors as recombinant vaccines. *Vaccine*, **10**, 8–9.

Baylor, E.R., Baylor, M., Blanchard, D.C., Syzdek, L.D. and Appel, C. (1977) Virus transfer from surf to wind. *Science*, **198**, 575–80.

Begon, M. and Harper, J.L. (1989) *Ecology: Individuals, Populations and Communities*. Blackwell Scientific Publications, Oxford.

Bennett, C.W. (1967a) Plant viruses: transmission by dodder, in *Methods in Virology*, Vol. 1, (eds K. Maramorosch and H. Koprowski), Academic Press, New York, pp. 393–401.

Bennett, C.W. (1967b) Epidemiology of leaf hopper-transmitted viruses. *Annu. Rev. Phytopathol.*, **5**, 87–108.

Bennett, C.W. (1969) Seed transmission of plant viruses. *Adv. Virus Res.*, **14**, 221–6.

Beran, G.W. (1981) *CRC Handbook Series in Zoonoses; Section B, Viral Zoonoses*, Vol. 1, CRC Press Inc., Boca Raton, Florida.

Berg, G. (1967) *Transmission of Viruses by the Water Route*, Interscience, New York.

Berger, P.H. and Zeyen, J.R. (1981) Extended aphid retention of MDMV: implications for long distance virus dispersal. *Phytopathology*, **71**, 203.

Bergion, M. and Dales, S. (1971) Comparative observations on poxviruses of invertebrates and vertebrates, in *Comparative Virology*, (eds K. Maramorosch and E. Kurstak), Academic Press, New York, pp. 169–205.

Bergold, G.H., Aizawa, K., Smith, K., Steinhaus, E.A. and Vago, C. (1960) The present status of insect virus nomenclature and classification. *Int. Bull. Bacteriol. Nomencl. Taxon.*, **10**, 259–62.

Berns, K.I. and Hauswirth, W.W. (1979) Adeno-associated viruses. *Adv. Virus. Res.*, **25**, 407–49.

Betley, M.J., Borst, D.W. and Regassa, L.B. (1992) Staphylococcal enterotoxins, toxic shock syndrome toxin and streptococcal pyrogenic exotoxins – a comparative study of their molecular biology. *Chem. Immunol.*, **55**, 1–35.

Bickle,T.A. and Kruger, D.H. (1993) The biology of DNA restriction. *Microbiol. Rev.*, **57**, 434–50.

Biggs, P.M. (1978) The veterinary professor and an intensive poultry industry. *Vet. Rec.*, **103**, 251–5.

Bird, F.T. and Burk, J.M. (1961) Artificially disseminated virus as a factor controlling the European spruce sawfly, *Diprion hercyniae* (Htg.) in the absence of introduced parasites. *Can. Entomol.*, **93**, 228–38.

Bird, J. and Maramorosch, K. (1978) Viruses and virus diseases associated with whiteflies. *Adv. Virus Res.*, **22**, 55–110.

Bishop, D.H.L. (1986) Ambisense RNA genomes of arenaviruses and phleboviruses. *Adv. Virus Res.*, **31**, 1–51.

Bishop, D.H.L. and Shope, R.E. (1979) Bunyaviridae, in *Comprehensive Virology*, Vol. 14, (eds H. Fraenkel-Conrat and R. Wagner), Plenum Press, New York, pp. 1–156.

Bisseru, B. (1972) *Rabies*. William Heinemann Medical Books Ltd, London.

Bitton, G. (1980) *Introduction to Environmental Virology*, Wiley-Interscience, New York.

Black, F.L. (1966) Measles endemicity in insular populations: critical community size and its evolutionary implication. *J. Theor. Biol.*, **11**, 207–11.

Blissard, G.W. and Rohrmann, G.F. (1990) Baculovirus diversity and molecular biology. *Annu. Rev. Entomol.*, **35**, 127.

Blua, M.A., Perring, T.M. and Madore, M.A. (1994) Plant virus-induced changes in aphid population development and temporal fluctuations in plant nutrients. *J. Chem. Ecol.*, **20**, 691–707.

Boccardo, G., Lisa, V. and Luisoni, E. (1987) Cryptic viruses. *Adv. Virus Res.*, **31**, 171–214.

Bode, O. (1977) Fragen der Quarantäne bei Kartofflen. *Potato Res.*, **20**, 349.

Boman, H.G. and Hultmark, D. (1987) Cell-free immunity in insects. *Annu. Rev. Microbiol.*, **41**, 103–26.

Bonning, B.C., Hirst, M., Possee, R.D. and Hammock, B.D. (1992) Further development of a recombinant baculovirus insecticide expressing the enzyme juvenile hormone esterase from *Heliothis virescens*. *Insect Biochem. Mol. Biol.*, **22**, 453–8.

Bos, L. (1970) *Symptoms of Virus Diseases in Plants*, 2nd edn., Centre for Agricultural Publications and Documentations, Wageningen.

Boshell, J.M. (1969) Kyasanur forest disease: ecological considerations. *Am. J. Trop. Med. Hyg.*, **18**, 67–80.

Bourdin, D. and Lecoq, H. (1991) Evidence that heteroencapsidation between two potyviruses is involved in aphid transmission of a nonaphidtransmissible isolate from mixed infections. *Phytopathology*, **81**, 1459–64.

Brasier, C.M. (1990) The unexpected element: mycovirus involvement in the outcome of two recent pandemics, dutch elm disease and chestnut blight, in *Pests Pathogens and Plant Communities*, (eds J.J. Burdon and S.R. Leather), Blackwell Scientific Publications, Oxford, pp. 289–307.

Bratbak, G., Heldal, M., Norland, S. and Thingstad, T.F. (1990) Viruses as partners in spring bloom microbial trophodynamics. *Appl. Environ. Microbiol.*, **56**, 1400–5.

Briddon, R.W., Pinner, M.S., Stanley, J. and Markham, P.G. (1990) Geminivirus coat protein gene replacement alters insect specificity. *Virology*, **177**, 85–94.

Brisou, J. (1976) *An Environmental Sanitation Plan for the Mediterranean Sea Board*. Public Health Paper No. 62. World Health Organization. Geneva.

Brown, D.J.F. and Trudgill, D.J. (1983) Differential transmissibility of arabis mosaic and strains of strawberry latent ringspot viruses by three populations of *Xiphinema diversicaudatum* (Nematoda, Dorylaimida) from Scotland, Italy and France. *Revue de Nematologie*, **6**, 229–38.

Brown, D.J.F., Ploeg, A.T. and Robinson, D.J. (1989) A review of reported associations between Trichodorus and Paratrichodorus species (Nematoda: Trichodoridae) and tobraviruses with a description of laboratory methods for examining virus transmission by trichodorids. *Revue de Nematologie*, **12**, 235–41.

Brown, R.M. (1972) Algal viruses. *Adv. Virus Res.*, **17**, 243–77.

Bruenn, J.A. (1991) Relationships among positive strand and double strand RNA viruses as viewed through RNA dependent RNA polymerases. *Nucleic Acids Res.*, **19**, 217–26.

Brunt, A.A. and Richards, K.E. (1989) Biology and molecular biology of furoviruses. *Adv. Virus Res.*, **36**, 1–32.

Brunt, A., Crabtree, K. and Gibbs, A. (1990) *Viruses of Tropical Plants*. CAB International.

Bukrinskaya, A.G. (1982) Penetration of viral genetic material into host cells. *Adv. Virus Res.*, **27**, 141–204.

Byrne, D.N. and Bellows,T.S. (1991) Whitefly biology. *Annu. Rev. Entomol.*, **36**, 431–57.

Campbell, A. (1961) Conditions for the existence of bacteriophage. *Evolution*, **15**, 153–65.

Campbell, A. (1981) Evolutionary significance of accessory DNA elements in bacteria. *Annu. Rev. Microbiol.*, **35**, 55–83.

Candelier-Harvey, P. and Hull, R. (1993) Cucumber mosaic virus genome is encapsidated in alfalfa mosaic virus coat protein expressed in transgenic tobacco plants. *Transgenic Research*, **2**, 277–85.

Carter, W. (1973) *Insects in Relation to Plant Disease*, 2nd edn, Wiley-Interscience Inc., New York.

Chang, S.H., Lustig, S., Strauss, E.S. and Strauss, J.H. (1988) Western equine encephalitis virus is a recombinant virus. *Proc. Natl Acad. Sci. USA*, **85**, 5997–6001.

Chang, S.L. (1968) Water-borne viral infections and their prevention. *Bull. WHO.*, **38**, 401–14.

Chapman, H.C. (1974) Biological control of mosquito larvae. *Annu. Rev. Entomol.*, **19**, 33–59.

Chasey, D. (1994) Possible origin of rabbit haemorrhagic disease in the UK. *Vet. Rec.*, **135**, 496–9.

Chen, J., MacFarlane, S.A. and Wilson, T.M.A. (1994) Detection and sequence analysis of a spontaneous deletion mutant of soil-borne wheat mosaic virus RNA-2 associated with increased symptom severity. *Virology*, **202**, 921–9.

Cheo, P.G. (1980) Antiviral factors in soil. *Soil Sci. Soc. Am. J.*, **44**, 62–7.

Chessin, M. (1972) Effect of radiation on viruses, in *Principles and Techniques in Plant Virology*, (eds C.I. Kado and H.O. Agrawal), Van Nostrand Reinhold, New York, pp. 531–45.

Citovsky, V. and Zambyski, P. (1993) Transport of nucleic acids through membrane channels: snaking through small holes. *Annu. Rev. Microbiol.*, **47**, 167–97.

Clark, C.S., Cleary, E.J., Schiff, G.M., Linnermanec, P., Phair, J.P. and Briggs, T.M. (1976) Disease risks of occupational exposure to sewage. *J. Environ. Eng. Div., Am. Soc. Civ. Eng.*, **102**, 375–588.

Clarke, D.K., Duarte, E.A., Moya, A., Elena, S.F., Domingo, E. and Holland, J. (1993) Genetic bottlenecks and population passages cause profound fitness differences in RNA viruses. *J. Virol.*, **67**, 222–8.

Cliff, A.D., Haggett, P.H., Ord, J.K. and Versey, G.R. (1981) *Spatial Diffusion*, Cambridge University Press, Cambridge.

Collins, C.H. (1983) *Laboratory-acquired Infections*, Butterworths, London.

Cooper, J.I. (1971) The distribution in Scotland of tobacco rattle virus and its nematode vectors in relation to soil type. *Plant Pathol.*, **20**, 51–8.

Cooper, J.I. (1979) *Virus Diseases of Trees and Shrubs*, Institute of Terrestrial Ecology, Cambridge.

Cooper, J.I. (1993) *Virus Diseases of Trees and Shrubs*, Chapman & Hall, London.

Cooper, J.I. and Asher, M.J.C. (1988) *Viruses with Fungal Vectors*, Association of Applied Biologists, Wellesbourne, UK.

Cooper, J.I. and Edwards, M.L. (1981) The distribution of poplar mosaic virus in hybrid poplars and virus detection by ELISA. *Ann. Appl. Biol.*, **99**, 53–61.

Cooper, J.I. and Edwards, M.L. (1986) Variations and limitations of enzyme amplified immunoassays, in *Development and Application in Virus Testing*, (eds R.A.C. Jones and L. Torrance), Association of Applied Biologists, Wellesbourne, Warks, pp. 139–54.

Cooper, J.I. and Harrison, B.D. (1973) The role of weed hosts and the distribution and activity of vectornematodes in the ecology of tobacco rattle virus. *Ann. Appl. Biol.*, **73**, 53–66.

Cooper, J.I. and Jones, A.T. (1983) Responses of plants to viruses: proposals for the use of terms. *Phytopathology*, **73**, 127–8.

Cooper, J.I., Jones, R.A.C. and Harrison, B.D. (1976) Field and glasshouse experiments on the control of potato mop-top virus. *Ann. Appl. Biol.*, **83**, 215–30.

Cooper, J.I., Massalski, P.R. and Edwards, M.L. (1984) Cherry leaf roll virus in the female gametophyte and seed of birch and its relevance to vertical virus transmission. *Ann. Appl. Biol.*, **105**, 55–64.

Cooper, J.I., Kelley, S.E. and Massalski, P.R. (1988) Virus–pollen interactions. *Adv. Dis. Vector Res.*, **5**, 221–49.

Cooper, J.I., Edwards, M.L., Rosenwasser, O. and Scott, N.W. (1994) Transgenic resistance genes from nepoviruses: efficacy and other properties. *N Z J. Crop Horticult. Sci.*, **22**, 129–37.

Cory, J.S., Hirst, M.L., Williams, T. *et al.* (1994) Field trial of a genetically improved baculovirus insecticide. *Nature*, **370**, 138–40.

Costa, A.S., Duffus, J.E. and Bardin, R. (1959) Malva yellows, an aphid transmitted virus disease. *J. Am. Soc. Sug. Beet Tech.*, **10**, 371–93.

Couch, J.A. and Courtney, L. (1977) Interaction of chemical pollutants and virus in a crustacean: a novel bioassay system. *Ann. N.Y. Acad. Sci.*, **298**, 497–504.

Crawford, A.M. (1989) Engineering of an *Oryctes* baculovirus recombinant: insertion of the polyhedrin gene from *Autographa californica* nuclear polyhedrosis virus. *J. Gen. Virol.*, **70**, 1017–24.

Creamer, R. and Falk, B.W. (1990) Direct detection of transcapsidated barley yellow dwarf luteovirus in doubly infected plants. *J. Gen. Virol.*, **71**, 211–17.

Croen, K.D. and Straus, S.E. (1991) Varicella-zoster virus latency. *Annu. Rev. Microbiol.*, **45**, 265–82.

Croghan, D.L., Matchett, A. and Koski, T.A. (1973) Isolation of porcine parvovirus from commercial trypsin. *Appl. Microbiol.*, **26**, 431–3.

D'Herelle, F. (1926) *The Bacteriophage and its Behaviour*, Bailliere, Tyndall and Cox, London.

David, W.A.L. (1975) The status of viruses pathogenic for insects and mites. *Annu. Rev. Entomol.*, **20**, 97–117.

David, W.A.L. (1978) The granulosis virus of *Pieris brassicae* (L.) and its relationship with its hosts. *Adv. Virus Res.*, **22**, 111–61.

Davidson, E.W. (1981) *Pathogenesis of Invertebrate Microbial Diseases*, Allanheld, Osmun, Ottowa, New Jersey.

Davidson, W.R., Kellog, F.E. and Doster, G.L. (1980) An epornitic of avian pox in wild bobwhite quail. *J. Wildl. Dis.*, **16**, 293–8.

Davies, J.W., Anderson, R.C., Karstad, L.H. and Trainer, D.O. (eds) (1971) *Infectious and Parasitic Diseases of Wild Birds*. Iowa State University Press, Ames, Iowa.

Davies, J.W., Karstad, L.H. and Trainer, D.O. (eds) (1981) *Infectious and Parasitic Diseases of Wild Mammals*, 2nd edn, Iowa State University Press, Ames, Iowa.

De Jong, W. and Ahlquist, P. (1992) A hybrid plant RNA virus made by transferring the noncapsid movement protein from a rod-shaped to an icosohedral virus is competent for systemic infection. *Proc. Natl Acad. Sci. USA*, **89**, 6808–12.

Derbyshire, J.B. and Brown, E.G. (1978) Isolation of animal viruses from farm livestock waste, soil and water. *J. Hyg. (Camb)*, **91**, 295–302.

Dessens, J.T., Nguyen, M. and Meyer, M. (1995) Primary structure and sequence analysis of RNA2 of a mechanically transmitted barley mild mosaic virus isolate: an evolutionary relationship between bymo- and furoviruses. *Arch. Virol.* **140**, 325–33.

Diamond, L.S. and Mattern, C.F.T. (1976) Protozoal viruses. *Adv. Virus Res.*, **20**, 87–112.

Diener, T.O. (1979) *Viroids and Viroid Diseases*, John Wiley and Sons, New York.

Dienstag, J.L., Gust, I.D., Lucus, C.R., Wong, D.C. and Purcell, R.H. (1976) Mussel associated with hepatitis type A, serological confirmation. *Lancet*, **i**, 561–3.

Dinant, S., Blaise, F., Kusiak, C., Astier-Manifacier, S. and Albouy, J. (1993) Heterologous resistance to potato virus Y in transgenic tobacco plants expressing the coat protein gene of lettuce mosaic potyvirus. *Phytopathology*, **83**, 818–24.

Dodds, J.A. and Hamilton, R.I. (1976) Structural interactions beween viruses as a consequence of mixed infections. *Adv. Virus Res.*, **20**, 33–86.

Doherty, P.C., Allan, W. and Eichenberger, M. (1992) Roles of alpha, beta and gamma, delta subsets in viral immmunity. *Annu. Rev. Immunol.*, **10**, 123–51.

Donaldson, A.I. and Ferris, N.P. (1975) The survival of foot-and-mouth disease virus in open air conditions. *J. Hyg. (Camb)*, **74**, 409–16.

Donaldson, A.I., Gloster, J., Harvey, L.D.J. and Deans, D.H. (1982) Use of prediction models to forecast and analyse airborne spread during the foot-and-mouth disease outbreaks in Brittany, Jersey and the Isle of Wight in 1981. *Vet. Record*, **110**, 53–7.

Done, S.H. and Paton, D.J. (1995) Porcine reproductive and respiratory syndrome: clinical disease, pathology and immunosuppression. *Vet. Rec.*, **136**, 32–5.

Duffus, J.E. (1963) Possible multiplication in the aphid vector of sowthistle yellow vein virus, a virus with an extremely long insect latent period. *Virology*, **21**, 194–202.

Dunn, P.E. (1986) Biochemical aspects of insect immunology. *Annu. Rev. Entomol.*, **31**, 321–39.

Edlin, G. (1978) Alteration of *Escherichia coli* outer membrane proteins by prophages. A model for benevolent virus–cell interactions, in *Persistent Viruses*, (eds J. Stevens, G.J. Todaro and F.L. Fox), Academic Press, New York.

Elena, S.F., Dopazo, J., Flores, R., Diener, T.O. and Mayo, A. (1991) Phylogeny of viroids, viroid-like satellite RNAs, and the viroid-like domain of hepatitis S virus RNA. *Proc. Natl Acad. Sci. USA*, **88**, 5631–4.

Ellis, R.W. (1993) *Hepatitis B Vaccines in Clinical Practice.* Marcel Dekker Inc., New York.

Entwistle, P.F. (1972) *Pests of Cocoa*, Longman, London.

Entwistle, P.F., Adams, P.H.W. and Evans, H.F. (1977) Epizootiology of a nuclear polyhedrosis virus in European spruce sawfly (*Gilpinia hercyniae*): the status of birds as dispersal agents of the virus during the larval season. *J. Invertebr. Pathol.*, **29**, 354–60.

Entwistle, P.F., Adams, P.H.W., Evans, H.F. and Rivers, C.F. (1983) Epizootiology of a nuclear polyhedrosis virus (baculoviridae) in European spruce sawfly (*Gilpinia hercyniae*): spread of disease from small epicentres in comparison with spread of baculovirus diseases in other hosts. *J. Appl. Ecol.*, **20** (2), 473–89.

Erdos, G.W. (1981) Virus-like particles in the acrasid cellular slime mold *Guttulinopsis vulgaris. Mycologia*, **73**, 785–8.

Evans, A.S. (ed.) (1976) *Viral Infections of Humans. Epidemiology and Control*, John Wiley and Sons, London, New York.

Evans, A.S. (ed.) (1982) *Viral Infections of Humans*, 2nd edn, Plenum Medical Book Company, New York and London.

Evans, H.F. and Harrap, K.A. (1982) Persistence of insect viruses, in *Virus Persistence*, (eds B.H.W. Mahy, A.C. Minson and G.K. Darby), Cambridge University Press, Cambridge, pp. 57–96.

Ewald, P.W. (1983) Host-parasite relations, vectors and the evolution of disease severity. *Microbiol. Rev.*, **55**, 586–620.

FAO/OIE/WHO (1977) *Animal Health Year Book*, Food and Agriculture Organisation of the United Nations, Rome.

Fenner, F. (1948) The epizootic behaviour of mouse pox (infectious ectromelia). *Br. J. Exp. Pathol.*, **29**, 69–91.

Fenner, F., MacAuslan, B.R., Mims, C.A., Sambrook, J. and White, D.O. (1974) *The Biology of Animal Viruses*, 2nd edn, Academic Press, New York and London.

Field, T.K., Patterson, C.A., Gergerich, R.C. and Kim, K.S. (1994) Fate of viruses in bean leaves after deposition by *Epilachna varivestis*, a beetle vector of plant viruses. *Phytopathology*, **84**, 1346–50.

Fields, B.N., Knipe, D.M., Chanock, R.M. *et al.* (1990) *Fields Virology*, 2nd Edition, Raven Press, New York.

Fine, P.E.M. (1975) Vectors and vertical transmission: an epidemiological perspective. *Ann. N.Y. Acad., Sci.*, **266**, 173–94.

Fitchen, J.H. and Beachy, R.N. (1993) Genetically engineered protection against viruses in transgenic plants. *Annu. Rev. Microbiol.*, **47**, 739–63.

Fleming, G. (1871) *Animal Plagues: their History, Nature and Prevention*, Chapman & Hall, London.

Fleming, G. (1882) *Animal Plagues: their History, Nature and Prevention*, Vol. 2, Ballière, Tindall and Cox, London.

Fleming, J.G.W. (1992) Polydnaviruses: mutualists and pathogens. *Annu. Rev. Entomol.*, **37**, 401–25.

Flewett, T.H. and Woode, G.N. (1978) The rotaviruses. *Arch. Virol.*, **57**, 1–23.

Foil, L.D. and Issel, C.J. (1991) Transmission of retroviruses by arthropods. *Annu. Rev. Entomol.*, **36**, 355–81.

Fraser, K.B. and Martin, S.J. (1978) *Measles Virus and its Biology*, Academic Press, London.

Fraser, R.S.S. (1990) The genetics of resistance to plant viruses. *Annu. Rev. Phytopathol.*, **28**, 179–200.

Frerichs, R.K. and Prawda, J. (1975) A computed simulation model for the control of rabies in an urban area of Colombia. *Manag. Sci.*, **22**, 411–21.

Freundt, E.A. (1974) The mycoplasmas, in *Bergey's Manual of Determinative Bacteriology*, 8th edn, (eds R.E. Buchanan and N.E. Gibbons), The Willams and Wilkins Co., Baltimore.

Fuxa, J.R. and Tanada, Y. (1987) *Epizootiology of Insect Diseases*, John Wiley & Sons Ltd., New York.

Gal, S., Pisan, B., Hohn, T., Grimsley, N. and Hohn, B. (1992) Agroinfection of transgenic plants leads to viable cauliflower mosaic virus by intermolecular recombination. *Virology*, **187**, 525–33.

Gameson, A.L.H. (1975) *Discharge of Sewage from Sea Outfalls*, Pergamon, Oxford.

Gause, G.F. (1934) *The Struggle for Existence*, Hafner, New York.

Gause, G.F. and Witt, A.A. (1935) Behaviour of mixed populations and the problem of natural selection. *Am. Nat.*, **69**, 596–609.

Geleziunas, R., Bour, S. and Wainberg, M.A. (1994) Human immunodeficiency virus type 1-associated CD4 down modulation. *Adv. Virus Res.*, **44**, 203–66.

Gerba, C.P. and Goyal, S.M. (1978) Detection and occurrence of enteric viruses in shellfish: a review. *J. Food Protec.*, **41**, 743–54.

Gergerich, R.C. and Scott, H.A. (1991) Determinants in the specificity of virus transmission by leaf-feeding beetles. *Adv. Dis. Vector Res.*, **8**, 1–13.

Gergerich, R.C., Scott, H.A. and Fulton, J.P. (1986) Evidence that ribonuclease in beetle regurgitant determines the transmission of plant viruses. *J. Gen. Virol.*, **67**, 367–70.

German, T.L., Ullman, D.E. and Moyer, J.W. (1992) Tospoviruses: diagnosis, molecular biology, physiology and vector relations. *Annu. Rev. Phytopathol.*, **30**, 315–48.

Gesteland, R.F. and Atkins, J.F. (1993) *The RNA World*, Cold Spring Harbor Laboratory Press, Cold Spring Harbor, New York.

Gibbs, A.J. (1969) Plant virus classification. *Adv. Virus Res.*, **14**, 263–328.

Gibbs, A.J. (ed.) (1973) *Viruses and Invertebrates*, North-Holland Publishing Co., Amsterdam.

Gibbs, A.J. (1980) A plant virus that partially protects its wild legume host against herbivores. *Intervirology*, **13**, 42–7.

Gibbs, A.J. and Harrison, B.D. (1976) *Plant Virology, the Principles*, Edward Arnold, London.

Gibbs, A.J., Gay, F.J. and Wetherly, A.H. (1970) A possible paralysis virus of termites. *Virology*, **40**, 1063–5.

Gibbs, E.P.J. (ed.) (1981) *Virus Diseases of Food Animals*, Vols I and II, Academic Press, London.

Gibbs, E.P.J., Taylor, W.P., Lawman, M.J.P. and Bryant, J. (1979) Classification of peste des petits ruminants virus as the fourth member of the genus Morbillivirus. *Intervirology*, **11**, 268–74.

Gibbs, M.J. and Cooper, J.I. (1995) A recombinational event in the history of luteoviruses probably induced by base-pairing between genomes of distinct viruses. *Virology*, **206**, 1129–32.

Gierer, A. and Schramm, G. (1956) Infectivity of ribonucleic acid from tobacco mosaic virus. *Nature*, **177**, 702–3.

Gildow, F.E. and D'arcy, C.J. (1988) Barley and oats as reservoirs for an aphid virus and the influence on barley yellow dwarf virus transmission. *Phytopathology*, **78**, 811–16.

Gillett, J.D. (1971) *Mosquitos*, Weidenfeld and Nicolson, London.

Goldbach, R., Le Gall, O. and Wellink, J. (1991) Alpha-like viruses in plants. *Semin. Virol.*, **2**, 19–25.

Gorman, O.T., Bean, W.J. and Webster, R.G. (1992) Evolutionary processes in influenza viruses: divergence, rapid evolution and stasis. *Curr. Top. Microbiol. Immunol.*, **176**, 75–97.

Govier, D.A., Kassanis, B. and Pirone, T.P. (1977) Partial purification and characterization of the potato virus Y helper component. *Virology*, **78**, 306–14.

Govindarajan, R. and Frederici, B.A. (1990) Ascovirus infectivity and effects of infection on the growth and development of Noctuid larvae. *J. Invertebr. Pathol.*, **56**, 291–9.

Greene, A.E. and Allison, R.F. (1994) Recombination between viral RNA and transgenic plant transcripts. *Science*, **264**, 1423.

Greenwood, M., Bradford-Hill, A. and Topley, W.W.C. (1936) *Experimental Epidemiology*. Special report No. 209. Medical Research Council. HMSO, London.

Gregg, N.M. (1941) Congenital cataract following German measles in the mother. *Trans. Ophthalmol. Soc. Austr.*, **3**, 35–46.

Griott, C., Gonzales-Scarano, F. and Nathanson, N. (1993) Molecular determinants of the virulence and infectivity of California serogroup bunyaviruses. *Annu. Rev. Microbiol.*, **45**, 117–38.

Guy, M.D., McIver, J.D. and Lewis, M.J. (1977) The removal of virus by a pilot treatment plant. *Water Res.*, **11**, 421–8.

Halstead, S.B. (1970) Observations related to pathogenesis of dengue hemorrhagic fever. *Yale J. Biol. Med.*, **42**, 350–62.

Hamilton, R.I., Edwardson, J.R., Francki, R.I.B. *et al.* (1981) Guidelines for the identification and characterization of plant viruses. *J. Gen. Virol.*, **54**, 223–41.

Hamilton, W.D. (1990) Sexual reproduction as an adaptation to resist parasites (a review). *Proc. Natl. Acad. Sci. USA*, **87**, 3566–73.

Hammond, J. (1981) Viruses occurring in Plantago species in England. *Plant Pathol.*, **30**, 237–43.

Hanada, K. and Harrison, B.D. (1977) Effects of virus genotype and temperature on seed transmission of nepoviruses. *Ann. Appl. Biol.*, **85**, 79–92.

Hansen, H.P. (1975) *Contribution to the Systemic Plant Virology. II. Codes, Data and Taxonomy*, Copenhagen, D.S.R. Forlag. The Royal Veterinary and Agricultural University.

Hardy, J.L., Houk, E.J., Kramer, L.D. and Reeves, W.C. (1983) Intrinsic factors affecting vector competence of mosquitoes for arboviruses. *Annu. Rev. Entomol.*, **28**, 229–62.

Hardy, V.G. and Teakle, D.S. (1992) Transmission of soybean mosaic virus by thrips in the presence and absence of virus-carrying pollen. *Ann. Appl. Biol.*, **121**, 315–20.

Harpaz, I. (1972) *Maize Rough Dwarf*, Israel University Press, Jerusalem. Harris, K.F. and Maramorosch, K. (eds) (1977) *Aphids as Virus Vectors*, Academic Press, New York, San Francisco and London.

Harris, K.F. and Maramorosch, K. (eds) (1977) *Aphids as virus vectors*, Academic Press, New York, San Francisco, London.

Harrison, B.D. (1992) Genetic engineering of virus resistance: a successful genetic alchemy. *Proc. R. Soc. Edinb.*, **99B**, 61–77.

Harrison, B.D. and Robinson, D.J. (1988) Molecular variation in vectorborne plant viruses: epidemiological significance. *Phil. Trans. R. Soc. Lond. B.*, **321**, 447–62.

Harwood, R.F. and James, M.T. (1979) *Entomology in Human and Animal Health*, 7th edn, Macmillan Publishing Co., Inc., New York.

Hassel, M.P., Comins, H.N. and May, R.M. (1991) Spacial structure and chaos in insect population dynamics. *Nature*, **353**, 255–8.

Hawker, L.E. and Linton, A.H. (1979) *Microorganisms Function, Form and Environment*, Edward Arnold, London.

Hayashiya, K. (1978) Red fluorescent protein in the digestive juice of the silkworm larvae fed on host-plant mulberry leaves. *Entomologia Experimentali et Applicata*, **24**, 428–36.

Hayes, C.G. and Wallis, R.C. (1977) Ecology of Western equine encephalomyelitis in the Eastern United States. *Adv. Virus Res.*, **21**, 37–83.

Hayes, W. (1964) *The Genetics of Bacteria and their Viruses*, Blackwell Scientific Publications, Oxford.

Henderson, H.M. and Cooper, J.I. (1977) Effect of thermal shock treatments on

References   181

symptom expression in test plants inoculated with potato aucuba mosaic virus. *Ann. Appl. Biol.*, **86**, 389–95.

Hershey, A.D. and Chase, M. (1952) Independent functions of viral protein and nucleic acid in growth of bacteriophage. *J. Gen. Physiol.*, **36**, 39–56.

Hewitt, W.B. and Chiarappa, L. (eds) (1977) Plant Health and Quarantine in International Transfer of Genetic Resources, CRC Press Inc., Cleveland, Ohio.

Hill, B.J. (1981) Virus diseases of fish, in *Virus Diseases in Food Animals*, Vol. 1, (ed. E.P.J. Gibbs), Academic Press, London, pp. 231–61.

Hobbs, H.A. and McLaughlin, M.R. (1990) A non aphid transmissible isolate of BYMV-Scott that is transmissible from mixed infection with pea mosaic virus. *Phytopathology*, **80**, 268–72.

Hoggan, M.D. (1971) Small DNA viruses, in *Comparative Virology*, (eds K. Maramorosch and E. Kurstak), Academic Press, New York.

Holland, J.J., Kennedy, I.T., Semler, B.L., Jones, C.L., Roux, L. and Grabau, E.A. (1980) Defective interfering RNA viruses and the host-cell response, in *Comprehensive Virology*, Vol. 16, (eds H. Fraenkel-Conrat and R.R. Wagner), Plenum Press, New York, pp. 137–92.

Holland, J.J., de la Torre, J.C. and Steinhauer, D. (1992) RNA virus populations as quasispecies. *Curr. Top. Microb. Immunol.*, **176**, 1–20.

Hollings, M. (1978) Mycoviruses: viruses that infect fungi. *Adv. Virus Res.*, **22**, 1–53.

Hoogstraal, H. (1981) Changing patterns of tick borne diseases in modern society. *Annu. Rev. Entomol.*, **26**, 75–99.

Hopkins, D.L. (1977) Diseases caused by leafhopper-borne rickettsia-like bacteria. *Annu. Rev. Phytopathol.*, **15**, 277–94.

Huang, A.S. and Baltimore, D. (1977) Defective interfering animal viruses, in *Comprehensive Virology*, Vol. 10, (eds H. Frankel-Conrat and R.R. Wagner), Plenum Press, New York, pp. 73–116.

Hugh-Jones, M.E. (1972) Epidemiological studies on the 1967/68 foot-and-mouth epidemic: attack rates and cattle density. *Res. Vet. Sci.*, **13**, 411–17.

Hugh-Jones, M.E., Allan, W.H., Dark, F.A. and Harper, G.J. (1973) The evidence for the air borne spread of Newcastle disease. *J. Hyg. (Camb)*, **71**, 325–39.

Hughes, J.d'A. and Ollenu, L.A.A. (1994) Mild strain protection of cocoa in Ghana against cocoa swollen shoot virus – a review. *Plant Pathol.*, **43**, 442–57.

Hull, R.N. (1968) in *The Simian Viruses*, Virology Mographs No. 2, Springer-Verlag, Wien and New York, pp. 1–66.

Jacobsen, G.S., Pearson, J.E. and Yuill, T.M. (1976) An epidemic of duck plague on a Wisconsin game farm. *J. Wildl. Dis.*, **12**, 20.

Jaques, R.P. (1967) The persistence of a nuclear polyhedrosis virus in the habitat of the host insect *Trichoplusia ni*. I. Polyhedra deposited on foliage. *Can. Entomol.*, **99**, 785–94.

Jaques, R.P. (1972) The inactivation of foliar deposits of viruses of *Trichoplusia ni Lepidoptera*: Noctuidae and *Pieris rapae Lepidoptera*: Pieridae and test on protectant additives. *Can. Entomol.*, **104**, 1985–94.

Jaques, R.P. (1975) Persistence, accumulation and denaturation of nuclear polyhedrosis and granulosis viruses, in *Baculovirus for Insect Pest Control*:

Safety Considerations, (eds M. Summers, R. Engler, L.A. Falcon and P. Vail), American Society for Microbiology, Washington, DC.

Johnson, K.M., Tesh, R.B. and Perakla, P.H. (1969) Epidemiology of vesicular stomatis virus: some new data and a hypothesis for transmission of the Indiana serotype. *J. Am. Vet. Med. Assoc.*, **155**, 2133–40.

Jones, A.T., McElroy, F.D. and Brown, D.J.F. (1981) Tests for transmission of cherry leaf roll virus using *Longidorus, Paralongidorus and Xiphinema* Nematodes. *Ann. Appl. Biol.*, **99**, 143–50.

Jones, D.L., Davies, C.R., Steele, G.M. and Nuttall, P.A. (1987) A novel mode of arbovirus transmission involving a nonviraemic host. *Science*, **237**, 775–7.

Jones, D.L., Davies, C.R., Williams, T., Cory, J. and Nuttall, P.A. (1990) Non-viraemic transmission of Thogoto virus: vector efficiency of *Rhipicephalus appendiculatus* and *Amblyomma variegatum*. *Trans. R. Soc. Trop. Med. Hyg.*, **84**, 846–8.

Joshi, R.L., Joshi, V. and Ow, D.W. (1990) BSMV genome mediated expression of a foreign gene in dicot and monocot plant cells. *EMBO J.*, **9**, 2663–9.

Kado, C.I. and Agrawal, H.O. (eds) (1972) *Principles and Techniques in Plant Virology*, Van Nostrand Reinhold, New York.

Kalmakoff, J., Williams, B.R.G. and Austin, J.F. (1977) Antiviral response in insects. *J. Invertebr. Pathol.*, **29**, 44–9.

Kamer, G. and Argos, P. (1984) Primary structural comparison of RNA-dependent polymerases from plant, animal and bacterial viruses. *Nucleic Acids Res.*, **12**, 7269–83.

Katzenelson, E., Buium, I. and Shuval, H.I. (1976) Risk of communicable disease infection associated with waste water irrigation in agricultural settlements. *Science*, **194**, 944–6.

Kelley, S.E. (1993) Viruses and the advantage of sex in *Anthoxanthum odoratum*: a review. *Plant Species Biol.*, **8**, 217–23.

Kelley, S.E. (1994) Viral pathogens and the advantage of sex in the perennial grass *Anthoxanthum odoratum*. *Phil. Trans. R. Soc. Lond. B*, **346**, 295–302.

Kelly, D.C. and Robertson, J.S. (1973) Icosahedral cytoplasmic deoxyriboviruses. *J. Gen. Virol.*, **20**, 17–41.

Killick, H.J. (1990) Influence of droplet size, solar ultraviolet light and pro-tectants, and other factors on the efficacy of baculovirus sprays against *Panolis flammea* (Schiff.) (Lepidoptera: Noctuidae). *Crop Protection*, **9**, 21–8.

King, L.A. and Possee, R.D. (1992) *The Baculovirus Expression System. A Labor-atory Guide*. Chapman & Hall, London.

Kirkegaard, K. and Baltimore, D. (1986) The mechanism of RNA recombination in poliovirus. *Cell*, **47**, 433–43.

Koenig, R. (1986) Plant viruses in rivers and lakes. *Adv. Virus Res.*, **31**, 321–33.

Konowalchuk, J. and Spiers, J.I. (1976) Virus inactivation by grapes and wines. *Appl. Environ. Microbiol.*, **32**, 757–63.

Kreiah, S., Strunk, G. and Cooper, J.I. (1994) Sequence analysis and location of capsid proteins within RNA2 of strawberry latent ringspot virus. *J. Gen. Virol.*, **75**, 2527–32.

Kurstak, E., Belloncik, S. and Brailowsky, C. (1969) Transformation de cellules de souris par un virus d'invertebres: le virus de la densonucleose (VDN). *C.R. Hebd. Seances Acad. Sci.*, **269**, 1716–19.

Kuwata, S., Masuta, C. and Takanami, Y. (1991) Reciprocal phenotype alterations between two satellite RNAs of cucumber mosaic virus. *J. Gen. Virol.*, **72**, 2385–9.

Lacey, R.W. (1994) *Mad cow disease: the history of BSE in Britain*, Cypsela Publications Ltd., St Helier, Jersey.

Lai, M.M.C. (1992) RNA recombination in animal and plant viruses. *Microbiol. Rev.*, **56**, 61–79.

Lamberti, F., Taylor, C.E. and Seinhorst, J.W. (eds) (1975) *Nematode Vectors of Plant Viruses*, Plenum Press, London and New York.

Lancaster, J.E. (1966) *Newcastle disease – a review, 1926–1964* (Monograph No. 3), Canada Department of Agriculture, Ottawa.

Lancaster, J.E. and Alexander, D.J. (1975) *Newcastle Disease Virus and Spread*, Monograph No. 11, Canada Department of Agriculture, Ottawa.

Larkin, E.P., Tierney, J.T. and Sullivan, R. (1976) Persistence of virus on sewage-irrigated vegetables. *J. Environ. Eng. Divi. Am. Soc. Civ. Eng.*, **102**, 29–35.

Larkin, P.J., Brittell, R.I.S., Banks, P. *et al.* (1989) Identification, characterisation and utilisation of sources of resistance to barley yellow dwarf virus, in *Barley Yellow Dwarf Virus, the Yellow Plague of Cereals*, (ed. P.A. Burnett), CIMYMYT, Mexico, pp. 415–20.

Laver, W.G. and Webster, R.G. (1979) Ecology of influenza viruses in lower mammals and birds. *Br. Med. Bull.*, **35**, 29–33.

Lemke, P.A. (1979) *Viruses and Plasmids in Fungi*, Marcell Dekker Inc., New York.

Levin, B.R., Stuart, F.M. and Chao, L. (1977) Resource-limited growth, competition and predation: a model and experimental studies with bacteria and bacteriophage. *Am. Nat.*, **111**, 3–24.

Lewis, T. (1966) Artificial wind breaks and the distribution of turnip mild yellows virus and *Scaptomyza apicalis* (Diptera) in a turnip crop. *Ann. Appl. Biol.*, **58**, 371–6.

Lindbo, J.A. and Dougherty, W.G. (1992a) Pathogen-derived resistance to a potyvirus: immune and resistant phenotypes in transgenic tobacco expressing altered forms of a potyvirus coat protein nucleotide sequence. *Mol. Plant-Microbe Interact.*, **5**, 144–53

Lindbo, J.A. and Dougherty, W.G. (1992b) Untranslatable transcripts of the tobacco etch virus coat protein gene sequence can interfere with tobacco etch virus replication in transgenic plants and protoplasts. *Virology*, **189**, 725–33.

Lindsay, M.D.A., Coelen, R.J. and Mackenzie, J.S. (1993) Genetic heterogeneity among isolates of Ross River virus from different geographical regions. *J. Virol.*, **67**, 3576–85.

Liu, Y.Y. and Cooper, J.I. (1994) Satellites of plant viruses. *Rev. Plant Pathol.*, **73**, 371–87.

Lively, C.M., Craddock, C. and Vrijenhoek, R.C. (1990) Red queen hypothesis supported by parasitism in sexual and clonal fish. *Nature*, **344**, 864–6.

Longworth, J.F. (1978) Small isometric viruses of invertebrates. *Adv. Virus Res.*, **23**, 103–57.

Longworth, J.F., Robertson, J.S., Tinsley, T.W., Rowlands, D.J. and Brown, F. (1973) Reactions between an insect picornavirus and naturally occurring IgM antibodies in several mammalian species. *Nature*, **242**, 314–16.

Louie, R., Findley, W.R. and Knoke, J.K. (1976) Variation in resistance within corn inbred lines to infection by maize dwarf mosaic virus. *Plant Disease Reporter*, **60**, 838–42.

Lunger, P.D. and Clark, H.F. (1978) Reptilia-related viruses. *Adv. Virus Res.*, **23**, 159–204.

Lycke, E., Blomberg, J., Berg, G., Eriksson, A. and Madsen, L. (1978) Epidemic acute diarrhoea in adults associated with infantile gastroenteritis virus. *Lancet*, **ii**, 1056–7.

MacClement, W.D. and Richards, M.G. (1956) Virus in wild plants. *Can. J. Bot.*, **34**, 793–9.

Maeda, S. (1989) Increased insecticidal effect by a recombinant baculovirus carrying a synthetic diuretic hormone gene. *Biochem. Biophys. Res. Commun.* **165**, 710–18.

Malipatil, M.B., Postle, A.C., Osmelek, J.A., Hill, M. and Moran, J. (1993) First record of *Frankiniella occidentalis* (Pergandee) in Australia (Thysanoptera: Thripidae). *J. Aust. Ent. Soc.*, **32**, 378.

Maramorosch, K. and Harris, K.F. (eds) (1979) *Leafhopper Vectors and Plant Disease Agents*, Academic Press, New York, San Francisco and London.

Maramorosch, K. and Harris, K.F. (eds) (1981) *Plant Diseases and Vectors: Ecology and Epidemiology*, Academic Press, New York, San Francisco and London.

Martelli, G.P. (1978) Nematode-borne viruses of grapevine, in *Plant Disease Epidemiology*, (eds P.R. Scott and A. Bainbridge), Blackwell Scientific Publications, Oxford, pp. 275–82.

Martignoni, M.E. and Iwai, P.J. (1975) A catalogue of viral diseases of insects and mites. *USDA Forest Service Technical Report*, PNW-40.

Martyn, E.B. (1971) *Plant Virus Names* (Phytopathological Papers No. 9), Commonwealth Mycological Institute, Kew.

Massalski, P.R. and Cooper, J.I. (1984) The location of virus like particles in the male gametophyte of birch, walnut and cherry naturally infected with cherry leaf roll virus and its relevance to vertical transmission of the virus. *Plant Pathology*, **33**, 255–62.

Matthews, R.E.F. (1980) Host plant responses to virus infection, in *Comprehensive Virology*, Vol. 16, (eds H. Fraenkel-Conrat and R.R. Wagner), Plenum Press, New York, pp. 297–359.

Matthews, R.E.F. (1981) *Plant Virology*, 2nd edn, Academic Press, New York.

Matthews, R.E.F. (1982) Fourth Report of the International Committee on Taxonomy of Viruses. Classification and nomenclature of viruses. *Intervirology*, **17**, 1–200.

Mayo, M.A. and Harrap, K.A. (1984) *Vectors in Virus Biology*, Academic Press, London.

Mayo, M.A. and Jolly, C.A. (1991) The 5'-terminal sequence of potato leafroll virus RNA: evidence of recombination between virus and host RNA. *J. Gen. Virol.*, **72**, 2591–5.

Mayo, M.A. and Martelli, G.P. (1993) New families and genera of plant viruses. *Arch. Virol.*, **133**, 496–8.

McIntosh, A.H. and Shamy, R. (1980) Biological studies of a baculovirus in a mammalian cell line. *Intervirology*, **13**, 331–41.

McLintock, J. (1978) Mosquito–virus relationships of American encephalitides. *Annu. Rev. Entomol.*, **23**, 17–37.

Merryweather, A.T., Weyer, U., Harris, M.P.G., Hirst, M. , Booth, T. and

Possee, R.D. (1990) Construction of genetically engineered baculovirus insecticides containing the *Bacillus thuringiensis* subsp. *kurstaki* HD-73 delta endotoxin. *J. Gen. Virol.*, **71**, 1535–44.

Messieha, M. (1969) Transmission of tobacco ringspot virus by thrips. *Phytopathology*, **59**, 943–5.

Meulenberg, J.J.M., Hulst, M.M., de Meijer, E.J. and Moonen, P.L.J.M. (1993) Lelystad virus, the causative agent of porcine epidemic abortion and respiratory syndrome (PEARS), is related to LDV and EHV. *Virology*, **198**, 62–72.

Meyers, G., Rumenapf, T. and Thiel, H.J. (1989) Ubiquitin in a togavirus. *Nature*, **341**, 491.

Mims, C.A. (1982) *The Pathogenesis of Infectious Disease*, Academic Press, London.

Mink, G.I. (1983) Possible role of honey bees in long distance spread of prunus necrotic ringspot virus from California into Washington sweet cherry orchards, in *Plant Virus Epidemiology*, (eds R.G. Plumb and M.J. Thresh), Blackwell Scientific Publications, Oxford, pp. 85–91.

Mink, G.I. (1992) Ilar virus vectors. *Adv. Virus Res.*, **9**, 262–81.

Mink, G.I. (1993) Pollen and seed transmittted viruses and viroids. *Annu. Rev. Phytopathol.*, **31**, 375–402.

Monath, T.P. (1986) *The Arboviruses: Epidemiology and Ecology.* CRC Press Inc., Boca Raton, Florida.

Morrow, A.W. (1969) Concentration of the virus of foot and mouth disease by foam floatation. *Nature*, **222**, 489–90.

Morse, S.S. (1994) *The Evolutionary Biology of Viruses*, Raven Press, New York.

Moss, B. (1991) Vaccinia virus: a tool for research and vaccine development. *Science*, **252**, 1662–7.

Moulder, J.W. (1966) The relation of the psittacosis group (Chlamdiae) to bacteria and viruses. *Annu. Rev. Microbiol.*, **20**, 107–30.

Mowat, W.P. (1968) *Olpidium brassicae*: electrophoretic mobility of zoospores associated with their ability to transmit tobacco necrosis virus. *Virology*, **34**, 565–8.

Muller, D.G. and Stache, B. (1992) Worldwide occurrence of virus-infections in filamentous marine brown algae. *Heligolander Meersunters*, **46**, 1–8.

Muller, D.G., Kawai, H., Stache, B. and Lanka, S. (1990) A virus infection in the marine brown alga *Ectocapus siliculosus* (Phaeophyteae). *Bot. Acta*, **103**, 72–82.

Mullis, K.B. (1990) The unusual origins of the polymerase chain reaction. *Scientific American*, **263**, 56–65.

Murant, A.F. and Lister, R.M. (1967) Seed transmission in the ecology of nematode-borne viruses. *Ann. Appl. Biol.*, **55**, 227–37.

Murant, A.F. and Mayo, M.A. (1982) Satellites of plant viruses. *Annu. Rev, Phytopathol.*, **20**, 49–70.

Murphy, F.A., Fauquet, C.M., Bishop, D.H.L. *et al.* (eds) (1995) *Sixth Report of the International Committee on Taxonomy of Viruses*, Virus taxonomy. Classification and Nomenclature of Viruses. Springer Verlag, Vienna.

Murphy, W.H. and Syverton, J.T. (1958) Absorption and translocation of mammalian viruses by plants. II. Recovery and distribution of viruses in plants. *Virology*, **6**, 623–36.

Murphy, W.H., Eylar, O.R., Schmidt, E.L. and Syverton, J.T. (1958) Absorption and translocation of mammalian viruses by plants. I. Survival of mouse encephalomyelitis and poliomyelitis viruses in soil and plant root environment. *Virology*, **6**, 612–22.

Murray, J.D., Stanley, E.A. and Brown, D.L. (1986) On the spatial spread of rabies among foxes. *Proc. R. Soc. Lond. B.*, **229**, 111–50.

Nahmias, A.J. and Reanney, D.C. (1977) The evolution of viruses. *Annu. Rev. Ecol. Syst.*, **8**, 29–49.

Nathanson, N., Yorke, J.A., Pianigiana, G. and Martin, J. (1978) Requirements for perpetuation and eradication of viruses in populations, in *Persistent Viruses*, (eds J. Stevens, G.J. Todaro and F.L. Fox), Academic Press, New York, pp. 75–100.

Nault, L.R. and Ammar, E.D. (1989) Leafhopper and planthopper transmission of plant viruses. *Annu. Rev. Entomol.*, **34**, 503–29.

Needham, G.R. and Teel, P.D. (1991) Off-host physiological ecology of ixodid ticks. *Annu. Rev. Entomol.*, **36**, 659–81.

Neilson, M.M. and Elgee, D.E. (1968) The method and role of vertical transmission of a nucleopolyhedrosis virus in the European spruce sawfly, *Diprion hercyniae. J. Invetebr. Pathol.*, **12**, 132–9.

Nerurkar, V.R., Song, J.W., Song, K.J., Nagle, J.W., Hjelle, B., Jenison, B. and Yanagihara, R. (1994) Genetic evidence for a hantavirus enzootic in deer mice (*peromysius mariculatus*) captured a decade before the recocognition of hantavirus pulmonary syndrome. *Virology*, **204**, 563–8.

Norton, D.C. (1978) *Ecology of Plant Parasitic Nematodes*, Wiley-Interscience, New York.

Offit, P.A. (1994) Rotaviruses: immunologic determinants of protection against infection and disease. *Adv. Virus Res.*, **44**, 161–202.

Oliviera, A.R. and Ponsen, M.B. (1966) The development of a viral antigen in the haemocytes of *P. brassicae* inoculated with *Tipula* iridescent virus. *Neth. J. Plant Pathol.*, **92**, 259–64.

Padan, E. and Shilo, M. (1973) Cyanophages. Viruses attacking blue-green algae. *Bacteriol. Rev.*, **37**, 343–70.

Parrish, C.R., Aqudro, C.F., Strassheim, M.L., Evermann, J.F., Sgro, J.Y. and Mohammed, H.O. (1991) Rapid antigenic-type replacement and DNA sequence evolution of canine parvovirus. *J. Virol.*, **65**, 6544–52.

Payne, C.C. (1988) Pathogens for the control of insects: where next? *Phil. Trans. R. Soc. Lond. B.*, **318**, 225–48.

Payne, C.C. and Rivers, C.F. (1976) A provisional classification of cytoplasmic polyhedrosis viruses based on the sizes of the RNA genome segments. *J. Gen. Virol.*, **33**, 71–85.

Pelham, J. (1972) Strain–genotype interaction of tobacco mosaic virus in tomato. *Ann. Appl. Biol.*, **71**, 219–28.

Pelham, J., Fletcher, J.T. and Hawkins, J.H. (1970) The establishment of a new strain of tomato mosaic virus resulting from the use of resistant varieties of tomato. *Ann. Appl. Biol.*, **65**, 293–7.

Ploeg, A.T., Asjes, C.J. and Brown, D.J.F. (1991) Tobacco rattle serotypes and associated nematode vector species of Trichodoridae in the bulb growing areas of the Netherlands. *Neth. J. Plant Pathol.*, **97**, 311–19.

Ploeg, A.T., Brown, D.J.F. and Robinson, D.J. (1992) The association between species of Trichodorus and Paratrichodorus vector nematodes and serotype of tobacco rattle virus. *Ann. Appl. Biol.*, **121**, 619–30.

Ploeg, A.T., Mathis, A., Bol, J.F., Brown, D.J.F. and Robinson, D.J. (1993) Susceptibility of transgenic tobacco plants expressing tobacco rattle virus coat protein to nematode-transmitted and mechanically inoculated tobacco rattle virus. *J. Gen. Virol.*, **74**, 2709–15.

Poinar, G.O., Hess, R.T. and Cole, A. (1980) Replication of an iridovirus in a nematode (Mermithidae). *Intervirology*, **14**, 316–20.

Potkyrus, I. (1991) Gene transfer to plants: assessment of published approaches and results. *Annu. Rev. Plant Physiol. Plant Mol. Biol.*, **42**, 205–25.

Preston, R. (1994) *The Hot Zone*, Doubleday, London.

Primrose, S.B., Seeley, N.D. and Logan, K.B. (1981) Methods of the study of virus ecology, in *Methods in Microbial Ecology*, (eds R.G. Burns and J.H. Slater), Blackwell Scientific Publications, Oxford, pp. 66–83.

Pringle, C.R. (1991) The order Mononegavirales. *Arch. Virol.*, **117**, 137–40.

Proeseler, G. (1966) Beziehungen zwischen der Rubenblattwanze *Piesma quadratum* Fieb. und dem Ruben Krausel virus. *Phytopathol. Z.*, **56**, 191–237.

Prusiner, S.B. (1992) Molecular biology and genetics of neurogenerative diseases caused by prions. *Adv. Virus Res.*, **41**, 241–80.

Ptashne, M. (1992) *A Genetic Switch: Phage [lambda] and Higher Organisms*, Blackwell Scientific Publications, Cambridge, MA, USA.

Purcell, A.H. (1979) Leafhopper vectors of xylem-borne plant pathogens, in *Leafhopper Vectors and Plant Disease Agents*, (eds K. Maramorosch and K.F. Harris), Academic Press, New York, pp. 603–25.

Rast, A.T.B. (1972) M11-16, an artificial symptomless mutant of tobacco mosaic virus for seedling inoculation of tomato crops. *Neth. J. Plant Pathol.*, **78**, 110–12.

Reanney, D.C. (1978) Non-coding sequences in adaptive genetics, in *Persistent Viruses*, (eds J. Stevens, G.J. Todaro and F.L. Fox), Academic Press, New York, pp. 311–30.

Reanney, D.C. and Ackermann, H.W. (1982) Comparative biology and evolution of bacteriophages. *Adv. Virus. Res.*, **27**, 205–80.

Reeves, R.H., O'Hara, B.F., Pavan, W.J., Gearhart, J.D. and Haller, O. (1988) Genetic mapping of the mx influenza resistance gene within the mouse chromosome 16 that is homologous to human chromosome 21. *J. Virol.*, **62**, 4372–5.

Reichmann, J.L., Lain, S. and Garcia, J.A. (1992) Highlights and prospects of potyvirus molecular biology *J. Gen. Virol.*, **73**, 1–16.

Reiter, W.D., Zillig, W. and Palm, P. (1988) Archaebacterial viruses. *Adv. Virus Res.*, **34**, 143–88.

Renaudin, J. and Bove, J.M. (1994) Spvi and Spv4, spiroplasma viruses with circular, single stranded DNA genomes, and their contribution to the molecular biology of spiroplasmas. *Adv. Virus Res.*, **44**, 429–63.

Rivers, T.M. and Horsfall, F.L. (1959) *Viral and Rickettsial Infections of Man*, Pitman, London.

Rochon, D., Kelly, R. and Siegel, A. (1986) Encapsidation of 18s rRNA by tobacco mosaic virus coat protein. *Virology*, **150**, 140–8.

Rochow, W.F. (1972) The role of mixed infections in the transmission of plant viruses by aphids. *Annu. Rev. Phytopathol.*, **10**, 101–24.

Rohrmann, G.F., Pearson, M.N., Bailey, J.T., Becker, R.R. and Beaudreau, G.S. (1981) N-terminal polyhedrin sequences and occluded baculovirus evolution. *J. Mol. Evol.*, **17**, 329–33.

Roitt, I.M. and Delves, P.J. (1992) *Encyclopedia of Immunology*, Academic Press, London.

Roizman, B., Desrosiers, R.C., Fleckenstein, B., Lopez, C., Minson, A.C. and Studdert, M.J. (1992) The family Herpesviridae: an update. *Intervirology*, **16**, 201–17.

Rose, D.J.W. (1978) Epidemiology of maize streak disease. *Annu. Rev. Entomol.*, **23**, 259–82.

Rothine, H.M., Chapdelaine, Y. and Hohn, T. (1994) Pararetroviruses and retroviruses: a comparative review of viral structure and gene expression strategies. *Adv. Virus Res.*, **44**, 1–67.

Rowson, K.E.K., Rees, T.A.L. and Mahy, B.H.J. (1981) *A Dictionary of Virology*, Blackwell Scientific Publications, Oxford.

Sanford, J.C. and Johnston, S.A. (1985) The concept of parasite-derived resistance-deriving resistance genes from the parasite's own genome. *J. Theoret. Biol.*, **113**, 395–405.

Sauve, G.J., Saragovi, H.U. and Greene, M.I. (1993) Reovirus receptors. *Adv. Virus Res.*, **42**, 325–41.

Scherer, W.F., Verna, J.E. and Richter, G.W. (1968) Nodamura virus, an ether- and chloroform-resistant arbovirus from Japan. *Am. J. Trop. Med. Hyg.*, **17**, 120–8.

Schleper, C., Kubo, K. and Zillig, W. (1992) The particle SSV1 from the extremely thermophilic archaeon Sulpholobus is a virus: demonstration of infectivity and of transfection of virus DNA. *Proc. Natl Acad. Sci. USA*, **89**, 7645–9.

Schlesinger, R.W. (ed.) (1980) *The Togaviruses*, Academic Press, New York.

Schrijnwerkers, C.C.F.M., Huijberts, N. and Bos, L. (1991) Zuccini yellow mosaic virus; two outbreaks in the Netherlands and seed transmissibility. *Neth. J. Plant Pathol.*, **97**, 187–91.

Schultz, U., Fitch, W.M., Ludwig, S., Mandler, J. and Scholtissek, C. (1991) Evolution of pig influenza viruses. *Virology*, **147**, 287–94.

Sdoodee, R. and Teakle, D.S. (1987) Transmission of tobacco streak virus by Thrips tabaci: a new method of plant virus transmission. *Plant Pathol.*, **36**, 377–80.

Sela, I. (1981) Plant virus interaction related to resistance and localisation of viral infections. *Adv. Virus. Res.*, **26**, 201–37.

Sellers, R.F. (1980) Weather, host and vector – their interplay in the spread of insect-borne animal virus diseases. *J. Hyg. (Camb)*, **85**, 65–102.

Shope, R.E. (1955) Epizootiology of virus diseases. *Adv. Vet. Sci.*, **2**, 1–46.

Shu, L.L., Lin, Y.P., Wright, S.M., Shortridge, K.F. and Webster, R.G. (1994) Evidence for interspecies transmission and reassortment of influenza A viruses in pigs in Southern China. *Virology*, **202**, 825–33.

Shukla, D., Ward, C.W. and Brunt, A.A. (1994) *The Potyviridae*, CAB International, Wallingford, Oxon.

Simmonds, J.H. (1959) Mild strain protection as a means of reducing losses from the Queensland woodiness virus in the passion vine. *Qd. J. Agric. Sci.*, **16**, 371–80.

Sleat, D.E. and Palukaitis, P. (1990) Site-directed mutagenesis of a plant virus satellite RNA changes its phenotype from ameliorative to necrogenic. *Proc. Natl Acad. Sci. USA*, **87**, 2946–50.

Smith, A.W., Vedros, N.A., Akers, T.F. and Gilmartin, W.G. (1978) Hazards of disease transfer from marine mammals to land animals: review and recent findings. *J. Am. Vet. Med. Assoc.*, **173**, 1131–3.

Smith, H.W. and Huggins, M.B. (1982) Successful treatment of experimental *Escherichia coli* infections in mice using bacteriophages: its general superiority over antibiotics. *J. Gen. Microbiol.*, **128**, 301–18.

Smith, H.W. and Huggins, M.B. (1983) Effectiveness of phages in treating experimental *Escherichia coli* diarrhoea in calves, piglets and lambs. *J. Gen. Microbiol.*, **129**, 2659–75.

Smith, H.W., Huggins, M.B. and Shaw, K.M. (1987) The control of experimental *Escherichia coli* diarrhoea in calves by means of bacteriophage. *J. Gen. Microbiol.*, **133**, 1111–26.

Smith, J.S. (1989) Rabies virus epitopic variation: use in ecologic studies. *Adv. Virus Res.*, **36**, 215–53.

Smith, K.M. (1972) *A Text Book of Plant Virus Diseases*, 3rd edn, Longman Group Ltd, London.

Spencer, D.C. and Price, R.W. (1992) Human immunodeficiency virus and the central nervous system. *Annu. Rev. Microbiol.*, **46**, 655–93.

Stent, G.S. (1963) *Molecular Biology of Bacterial Viruses*, Freeman, San Francisco.

Stewart, L.M.D., Hirst, M., Ferber, M.L. *et al.* (1991) Construction of an improved baculovirus insecticide containing an insect-specific toxin gene. *Nature*, **352**, 85–8.

Stoltz, D.B. and Vinson, S.B. (1979) Viruses and parasitism in insects. *Adv. Virus Res.*, **24**, 125–71.

Stortz, J. and Page, L.A. (1971) Taxonomy of the Chlamydiae. *Int. J. Syst. Bacteriol.*, **21**, 332–4.

Strode, G.K. (1951) *Yellow Fever*, McGraw-Hill Book Company Ltd, New York, Toronto and New York.

Stuart Harris, C.H. and Schild, G.C. (1976) *Influenza, the Viruses and the Disease*, Edward Arnold, London.

Sylvester, E.S. (1969) Evidence of transovarial passage of sowthistle yellow vein virus in the aphid *Hyperomyzus lactucae*. *Virology*, **38**, 440–6.

Sylvester, E.S. (1980) Circulative and propagative virus transmission by aphids. *Annu. Rev. Entomol.*, **25**, 257–86.

Takahashi, Y. and Orlob, G.B. (1969) Distribution of wheat streak mosaic virus-like particle in *Aceria tulipae*. *Virology*, **38**, 230–40.

Taviadoraki, P., Benvenuto, E., Trinca, S., De Martinis, D., Cattaneo, A. and Galeffi, P. (1993) Transgenic plants expressing a functional single-chain Fv antibody are specifically protected from virus attack. *Nature*, **366**, 469–72.

Taylor, N.C. and Ghabriel, S.A. (1986) Breeding forage legumes for resistance to viruses, in *Viruses Infecting Forage Legumes*, Vol. III, Monograph No. 14, (eds J.R. Edwardson and R.G. Christie), University of Florida, Gainsville, pp. 609–23.

Thresh, J.M. (1976) Gradients of plant virus diseases. *Ann. Appl. Biol.*, **82**, 381–406.

Thresh, J.M. (1980a) The origins and epidemiology of some important plant virus diseases. *Ann. Appl. Biol.*, **5**, 1–65.

Thresh, J.M. (1980b) An ecological approach to the epidemiology of plant virus diseases, in *Comparative Epidemiology*, (eds J. Kranz and J. Palti), Pudoc, Wageningen.

Thresh, J.M. (ed.) (1981) *Pests, Pathogens and Vegetations*, Pitman, Boston, London and Melbourne.

Thresh, J.M. (1982) Cropping practices and virus spread. *Annu. Rev. Phytopathol.*, **20**, 193–218.

Tien, P. and Wu, G. (1991) Satellite RNA for the biocontrol of plant disease. *Adv. Virus Res.*, **39**, 321–39.

Tinsley, T.W. and Longworth, J.F. (1973) Parvoviruses. *J. Gen. Virol.*, **20**, 7–15.

Truve, E., Aaspollu, A., Honkanen, J. et al. (1993) Transgenic potato plants expressing mammalian 2–5 oligoadenylate synthetase are protected from potato virus X infection under field conditions. *Bio/Technology*, **11**, 1048–52.

Turek, L.P. (1994) The structure, function and regulation of papillomaviral genes in infection and cervical cancer. *Adv. Virus Res.*, **44**, 305–56.

van den Heuvel, J.F.J.M., Verbeek, M. and van der Wilk, F. (1994) Endosymbiotic bacteria associated with circulative transmission of potato leafroll virus by *Myzus persicae*. *J. Gen. Virol.* **75**, 2559–65.

van der Pijl, L. (1972) *Principles of Dispersal in Higher Plants*, Springer-Verlag, Berlin, Heidelberg and New York.

van der Plank, J.E. (1960) Analysis of epidemics, in *Plant Pathology*, Vol. 3, (eds J.G. Horsfall and A.E. Dimond), Academic Press, New York and London, pp. 230–89.

van der Vlugt, R.A.A. and Goldbach, R.W. (1993) Tobacco plants transformed with the potato virus Y^N coat protein gene are protected against different PVY isolates and against aphid-mediated infection. *Transgenic Research*, **2**, 109–14.

van Etten, J.L., Lane, L.C. and Meints, R.H. (1991) Viruses and virus-like particles of eukaryotic algae. *Microbiol. Rev.*, **55**, 586–620.

van Regenmortel, M.H.V. (1982) *Serology and Immunochemistry of Plant Viruses*, Academic Press, New York.

Viswanathan, R. (1957) Infectious hepatitis in New Delhi 1955–56. Epidemiology. *Indian J. Med. Res.*, **xlv**, Suppl. 1–29.

Walkey, D.G.A., Le Coq, H., Collier, R. And Dobson, S. (1992) Studies on the control of zuccini yellow mosaic virus in courgettes by mild strain protection. *Plant Pathol.*, **41**, 762–71,

Walters, H.J. (1969) Beetle transmission of plant viruses. *Adv. Virus Res.*, **15**, 339–63.

Wang, A.L. and Wang, C.C. (1991) Viruses of the protozoa. *Annu. Rev. Microbiol.*, **45**, 251–63.

Wang, D. and Maule, A.J. (1995) Inhibition of host gene expression associated with plant virus replication. *Science*, **267**, 229–31.

Wang, H.L., Gonsalves, D., Providenti, R. and Lecoq, H. (1991) Effectiveness of cross protection by a mild strain of zuccini yellow mosaic virus in cucumber, melon and squash. *Plant Disease*, **75**, 203–7.

Ward, R.L. and Ashley, C.S. (1978) Heat inactivation of enteric viruses in dewatered waste water sludge. *Appl. Environ. Microbiol.*, **36**, 898–905.

Warner, R.E. (1968) The role of introduced diseases in the extinction of the endemic Hawaiian avifauna. *Condor*, **70**, 101–20.

Watson, M.A. (1946) The transmission of beet mosaic and beet yellows viruses by aphids; a comparative study of a non-persistent and a persistent virus having host plants and vectors in common. *Proc. R. Soc. Ser. B*, **133**, 200–19.

Webster, R.G. and Granoff, A. (1994) *Encyclopedia of Virology*, Academic Press Ltd, London.

Webster, R.G., Yakhno, M., Hinshaw, V.S., Bean, W.J. and Murti, K.G. (1978) Intestinal influenza: replication and characterization of influenza viruses in ducks. *Virology*, **84**, 268–78.

Weir, D.M. (ed.) (1978) *Handbook of Experimental Immunology*, Blackwell Scientific Publications, Oxford.

Wellink, J. and van Kammen, A. (1989) Cell to cell transport of cowpea mosaic virus requires both the 58K/48K proteins and the capsid protein. *J. Gen. Virol.*, **70**, 2279–86.

Wen, F. and Lister, R.M. (1991) Heterologous encapsidation in mixed infections among four isolates of barley yellow dwarf virus. *J. Gen. Virol.*, **72**, 2217–23.

West, G.P. (1995) *Black's Veterinary Dictionary*, 18th edn, A. & C. Black Ltd, London.

Whitaker-Dowling, P., Maasaab, H.F. and Youngner, J.S. (1991) Dominant-negative mutants as antiviral agents: simultaneous infection with the cold-adapted live-virus vaccine for influenza protects ferrets from disease produced by wild-type influenza. *J. Infect. Dis.*, **164**, 1200–2.

Whitcomb, R.F. and Tully, J.G. (1979) *The Mycoplasmas*, Vol. III, Academic Press, New York.

White, J.L., Tousignant, M.E., Geletka, L.M. and Kaper, J.M. (1995) The replication of a necrogenic cucumber mosaic virus satellite is temperature sensitive in tomato. *Arch. Virol.*, **140**, 53–63.

Wilder, F.H. and Dardiri, A.H. (1978) San Miguel sea lion virus fed to mink and pigs. *Can. J. Comp. Med.*, **42**, 200–4.

Williams, T. (1994) Comparative studies of iridoviruses: further support for a new classification. *Virus Res.*, **33**, 99–121.

Williams, T. and Cory, J.S. (1994) Proposals for a new classification of irido-viruses. *J. Gen. Virol.*, **75**, 1291–301.

Wilson, T.M.A. (1984) Cotranslational disassembly increases the efficiency of expression of TMA RNA in wheat germ cell-free extracts. *Virology*, **138**, 353–6.

Wilson, T.M.A. (1993) Strategies to protect crop plants against viruses: pathogen-derived resistance blossoms. *Proc. Natl Acad. Sci. USA*, **90**, 3134–41.

Wood, H.A. and Granados, R.R. (1991) Genetically engineered baculoviruses as agents for pest control. *Annu. Rev. Microbiol.*, **45**, 69–87.

Work, T.H. (1958) Russian spring–summer virus in India. Kyasanur Forest disease. *Prog. Med. Virol.*, **1**, 248–79.

Yahara, T. and Oyama, K. (1993) Effects of virus infection on demographic traits of an angiospermous population of *Eupatorium chinense* (Asteraceae). *Oecologia*, **96**, 310–15.

Yewdell, J.W. and Gerhard, W. (1981) Antigenic characterization of viruses by monoclonal antibodies. *Annu. Rev. Microbiol.*, **351**, 185–206.

Yorke, J.A., Nathanson, N., Pianigiani, G. and Martin, J. (1979) Seasonality and the requirements for perpetuation and eradication of viruses in populations. *Am. J. Epidemiol.*, **109**, 103–23.

Young, E.C. (1974) The epizootiology of two pathogens of the coconut palm rhinoceros beetle. *J. Invertebr. Pathol.*, **24**, 82–92.

Zaccomer, B., Haenni, A.-L. and Macaya, G. (1995) The remarkable variety of plant RNA virus genomes. *J. Gen. Virol.*, 231–47.

Zanatto, P.M. de A., Gao, G.F., Gritsun, T. *et al.* (1995) An arbovirus cline across the Northern Hemisphere. *Virology*, **210**, 152–159

Zavada, J. (1982) The pseudotypic paradox. *J. Gen. Virol.*, **63**, 15–24

Zhang, Y., Burbank, D.E. and van Etten, J.L. (1988) Chlorella viruses isolated in China. *Appl. Environ. Microbiol.*, **54**, 2170–3.

Zuckerman, A.J. and Howard, C.R. (1979) *Hepatitis Viruses of Man*, Academic Press, London.

Author index

Subject index

Page numbers in *italics* refer to figures or tables.